Tales of Chinatown

Sax Rohmer

IAP © ~ 2009

IAP. Las Vegas, NV - USA.

Printed in California.

Rohmer, Sax.

Tales of Chinatown / Sax Rohmer – 1st ed.

   1. Literature

Book Cover Image
© Isabel Poulin | Dreamstime.com

CONTENTS

4

THE DAUGHTER OF HUANG CHOW

CHAPTER I

"DIAMOND FRED"

In the saloon bar of a public-house, situated only a few hundred yards from the official frontier of Chinatown, two men sat at a small table in a corner, engaged in earnest conversation. They afforded a sharp contrast. One was a thick-set and rather ruffianly looking fellow, not too cleanly in either person or clothing, and, amongst other evidences that at one time he had known the prize ring, possessing a badly broken nose. His companion was dressed with that spruceness which belongs to the successful East End Jew; he was cleanly shaven, of slight build, and alert in manner and address.

Having ordered and paid for two whiskies and sodas, the Jew, raising his glass, nodded to his companion and took a drink. The glitter of a magnificent diamond which he wore seemed to attract the other's attention almost hypnotically.

"Cheerio, Freddy!" said the thick-set man. "Any news?"

"Nothing much," returned the one addressed as Freddy, setting his glass upon the table and selecting a cigarette from a packet which he carried in his pocket.

"I'm not so sure," growled the other, watching him suspiciously. "You've been lying low for a long time, and it's not like you to slack off except when there's something big in sight."

"Hm!" said his companion, lighting his cigarette. "What do you mean exactly?"

Jim Poland--for such was the big man's name--growled and spat reflectively into a spittoon.

"I've had my eye on you, Freddy," he replied; "I've had my eye on you!"

"Oh, have you?" murmured the other. "But tell me what you mean!"

Beneath his suave manner lay a threat, and, indeed, Freddy Cohen, known to his associates as "Diamond Fred," was in many ways a formidable personality. He had brought to his chosen profession of crook a first-rate American training, together with all that mental agility and cleverness which belong to his race, and was at once an object of envy and admiration amongst the fraternity which keeps Scotland Yard busy.

Jim Poland, physically a more dangerous character, was not in the same class with him; but he was not without brains of a sort, and Cohen, although smiling agreeably, waited with some anxiety for his reply.

"I mean," growled Poland, "that you're not wasting your time with Lala Huang for nothing."

"Perhaps not," returned Cohen lightly. "She's a pretty girl; but what business is it of yours?"

"None at all. I ain't interested in 'er good looks; neither are you."

5

Cohen shrugged and raised his glass again.

"Come on," growled Poland, leaning across the table. "I know, and I'm in on it. D'ye hear me? I'm in on it. These are hard times, and we've got to stick together."

"Oh," said Cohen, "that's the game, is it?"

"That's the game right enough. You won't go wrong if you bring me in, even at fifty-fifty, because maybe I know things about old Huang that you don't know."

The Jew's expression changed subtly, and beneath his drooping lids he glanced aside at the speaker. Then:

"It's no promise," he said, "but what do you know?"

Poland bent farther over the table.

"Chinatown's being watched again. I heard this morning that Red Kerry was down here."

Cohen laughed.

"Red Kerry!" he echoed. "Red Kerry means nothing in my young life, Jim."

"Don't 'e?" returned Jim, snarling viciously. "The way he cleaned up that dope crowd awhile back seemed to show he was no jug, didn't it?"

The Jew made a racial gesture as if to dismiss the subject.

"All right," continued Poland. "Think that way if you like. But the patrols have been doubled. I suppose you know that? And it's a cert there are special men on duty, ever since the death of that Chink."

Cohen shifted uneasily, glancing about him in a furtive fashion.

"See what I mean?" continued the other. "Chinatown ain't healthy just now."

He finished his whisky at a draught, and, standing up, lurched heavily across to the counter. He returned with two more glasses. Then, reseating himself and bending forward again:

"There's one thing I reckon you don't know," he whispered in Cohen's ear. "I saw that Chink talking to Lala Huang only a week before the time he was hauled out of Limehouse Reach. I'm wondering, Diamond, if, with all your cleverness, you may not go the same way."

"Don't try to pull the creep stuff on me, Jim," said Cohen uneasily. "What are you driving at, anyway?"

"Well," replied Poland, sipping his whisky reflectively, "how did that Chink get into the river?"

"How the devil do I know?"

"And what killed him? It wasn't drowning, although he was all swelled up."

"See here, old pal," said Cohen. "I know 'Frisco better than you know Limehouse. Let me tell you that this little old Chinatown of yours is pie to me. You're trying to get me figuring on Chinese death traps, secret poisons, and all that junk. Boy, you're wasting your poetry.

Even if you did see the Chink with Lala, and I doubt it-- Oh, don't get excited, I'm speaking plain--there's no connection that I can see between the death of said Chink and old Huang Chow."

"Ain't there?" growled Poland huskily. He grasped the other's wrist as in a vise and bent forward so that his battered face was close to the pale countenance of the Jew. "I've been covering old Huang for months and months. Now I'm going to tell you something. Since the death of that Chink Red Kerry's been covering him, too."

"See here!" Cohen withdrew his arm from the other's grasp angrily. "You can't freeze me out of this claim with bogey stuff. You're listed, my lad, and you know it. Chief Inspector Kerry is your pet nightmare. But if he walked in here right now I could ask him to have a drink. I wouldn't but I could. You've got the wrong angle, Jim. Lala likes me fine, and although she doesn't say much, what she does say is straight. I'll ask her to-night about the Chink."

"Then you'll be a damned fool."

"What's that?"

"I say you'll be a damned fool. I'm warning you, Freddy. There are Chinks and Chinks. All the boys know old Huang Chow has got a regular gold mine buried somewhere under the floor. But all the boys don't know what I know, and it seems that you don't either."

"What is that?"

Jim Poland bent forward more urgently, again seizing Cohen's wrist, and:

"Huang Chow is a mighty big bug amongst the Chinese," he whispered, glancing cautiously about him. "He's hellish clever and rotten with money. A man like that wants handling. I'm not telling you what I know. But call it fifty-fifty and maybe you'll come out alive."

The brow of Diamond Fred displayed beads of perspiration, and with a blue silk handkerchief which he carried in his breast pocket he delicately dried his forehead.

"You're an old hand at this stuff, Jim," he muttered. "It amounts to this, I suppose; that if I don't agree you'll queer my game?"

Jim Poland's brow lowered and he clenched his fists formidably. Then:

"Listen," he said in his hoarse voice. "It ain't your claim any more than mine. You've covered it different, that's all. Yours was always the petticoat lay. Mine's slower but safer. Is anyone else in with you?"

"No."

"Then we'll double up. Now I'll tell you something. I was backing out."

"What? You were going to quit?"

"I was."

"Why?"

"Because the thing's too dead easy, and a thing like that always looks like hell to me."

Freddy Cohen finished his glass of whisky.

"Wait while I get some more drinks," he said.

In this way, then, at about the hour of ten on a stuffy autumn night, in the crowded bar of that Wapping public-house, these two made a compact; and of its outcome and of the next appearance of Cohen, the Jewish-American cracksman, within the ken of man, I shall now proceed to tell.

CHAPTER 2

THE END OF COHEN

"I've been expecting this," said Chief Inspector Kerry. He tilted his bowler hat farther forward over his brow and contemplated the ghastly exhibit which lay upon the slab of the mortuary. Two other police officers--one in uniform--were present, and they treated the celebrated Chief Inspector with the deference which he had not only earned but had always demanded from his subordinates.

Earmarked for important promotion, he was an interesting figure as he stood there in the gloomy, ill-lighted place, his pose that of an athlete about to perform a long jump, or perhaps, as it might have appeared to some, that of a dancing-master about to demonstrate a new step.

His close-cropped hair was brilliantly red, and so was his short, wiry, aggressive moustache. He was ruddy of complexion, and he looked out unblinkingly upon the world with a pair of steel-blue eyes. Neat he was to spruceness, and while of no more than medium height he had the shoulders of an acrobat.

The detective who stood beside him, by name John Durham, had one trait in common with his celebrated superior. This was a quick keenness, a sort of alert vitality, which showed in his eyes, and indeed in every line of his thin, clean-shaven face. Kerry had picked him out as the most promising junior in his department.

"Give me the particulars," said the Chief Inspector. "It isn't robbery. He's wearing a diamond ring worth two hundred pounds."

His diction was rapid and terse--so rapid as to create the impression that he bit off the ends of the longer words. He turned his fierce blue eyes upon the uniformed officer who stood at the end of the slab.

"They are very few, Chief Inspector," was the reply. "He was hauled out by the river police shortly after midnight, at the lower end of Limehouse Reach. He was alive then--they heard his cry--but he died while they were hauling him into the boat."

"Any statement?" rapped Kerry.

"He was past it, Chief Inspector. According to the report of the officer in charge, he mumbled something which sounded like: 'It has bitten me,' just before he became unconscious."

"'It has bitten me,'" murmured Kerry. "The divisional surgeon has seen him?"

"Yes, Chief Inspector. And in his opinion the man did not die from drowning, but from some form of virulent poisoning."

"Poisoning?"

"That's the idea. There will be a further examination, of course. Either a hypodermic injection or a bite."

"A bite?" said Kerry. "The bite of what?"

"That I cannot say, Chief Inspector. A venomous reptile, I suppose."

Kerry stared down critically at the swollen face of the victim, and then glanced sharply aside at Durham.

"Accounts for his appearance, I suppose," he murmured.

"Yes," said Durham quietly. "He hadn't been in the water long enough to look like that." He turned to the local officer. "Is there any theory as to the point at which he went in?"

"Well, an arrest has been made."

"By whom? of whom?" rapped Kerry.

"Two constables patrolling the Chinatown area arrested a man for suspicious loitering. He turned out to be a well-known criminal--Jim Poland, with a whole list of convictions against him. They're holding him at Limehouse Station, and the theory is that he was operating with..." He nodded in the direction of the body.

"Then who's the smart with the swollen face?" inquired Kerry. "He's a new one on me."

"Yes, but he's been identified by one of the K Division men. He is an American crook with a clean slate, so far as this side is concerned. Cohen is his name. And the idea seems to be that he went in at some point between where he was found by the river police and the point at which Jim Poland was arrested."

Kerry snapped his teeth together audibly, and:

"I'm open to learn," he said, "that the house of Huang Chow is within that area."

"It is."

"I thought so. He died the same way the Chinaman died awhile ago," snapped Kerry savagely.

"It looks very queer." He glanced aside at the local officer. "Cover him up," he ordered, and, turning, he walked briskly out of the mortuary, followed by Detective Durham.

Although dawn was not far off, this was the darkest hour of the night, so that even the sounds of dockland were muted and the riverside slept as deeply as the great port of London ever sleeps. Vague murmurings there were and distant clankings, with the hum of machinery which is never still.

Few of London's millions were awake at that hour, yet Scotland Yard was awake in the person of the fierce-eyed Chief Inspector and his subordinate. Perhaps those who lightly criticize the Metropolitan Force might have learned a new respect for the tireless vigilance which keeps London clean and wholesome, had they witnessed this scene on the borders of Limehouse, as Kerry, stepping into a waiting taxi-cab accompanied by Durham, proceeded to Limehouse Police Station in that still hour when the City slept.

The arrival of Kerry created something of a stir amongst the officials on duty. His reputation in these days was at least as great as that of the most garrulous Labor member.

The prisoner was in cells, but the Chief Inspector elected to interview him in the office; and accordingly, while the officer in charge sat at an extremely tidy writing-table, tapping the blotting-pad with a pencil, and Detective John Durham stood beside him, Kerry paced up and down the little room, deep in reflection, until the door opened and the prisoner was brought in.

One swift glance the Chief Inspector gave at the battle-scarred face, and recognized instantly that this was a badly frightened man. Crossing to the table he took up a typewritten slip which lay there, and:

"Your name is James Poland?" he said. "Four convictions; one, robbery with violence."

Jim Poland nodded sullenly.

"You were arrested at the corner of Pekin Street about midnight. What were you doing there?"

"Taking a walk."

"I'll say it again," rapped Kerry, fixing his fierce eyes upon the man's face. "What were you doing there?"

"I've told you."

"And I tell you you're a liar. Where did you leave the man Cohen?"

Poland blinked his small eyes, cleared his throat, and looked down at the floor uneasily. Then:

"Who's Cohen?" he grunted.

"You mean, who was Cohen?" cried Kerry.

The shot went home. The man clenched his fists and looked about the room from face to face.

"You don't tell me..." he began huskily.

"I've told you," said Kerry. "He's on the slab. Spit out the truth; it'll be good for your health."

The man hesitated, then looked up, his eyes half closed and a cunning expression upon his face.

"Make out your own case," he said. "You've got nothing against me."

Kerry snapped his teeth together viciously.

"I've told you what happened to your pal," he warned. "If you're a wise man you'll come in

10

on our side, before the same thing happens to you."

"I don't know what you're talking about," growled Poland.

Kerry nodded to the constable at the doorway.

"Take him back," he ordered.

Jim Poland being returned to his cell, Kerry, as the door closed behind the prisoner and his guard, stared across at Durham where he stood beside the table.

"An old hand," he said. "But there's another way." He glanced at the officer in charge. "Hold him till the morning. He'll prove useful."

From his waistcoat pocket he took out a slip of chewing gum, unwrapped it, and placed the mint-flavored wafer between his large white teeth. He bit upon it savagely, settled his hat upon his head, and, turning, walked toward the door. In the doorway he paused.

"Come with me, Durham," he said. "I am leaving the conduct of the case entirely in your hands from now onward."

Detective Durham looked surprised and not a little anxious.

"I am doing so for two reasons," continued the Chief Inspector. "These two reasons I shall now explain."

CHAPTER 3

THE SECRET TREASURE-HOUSE

Unlike its sister colony in New York, there are no show places in Limehouse. The visitor sees nothing but mean streets and dark doorways. The superficial inquirer comes away convinced that the romance of the Asiatic district has no existence outside the imaginations of writers of fiction. Yet here lies a secret quarter, as secret and as strange, in its smaller way, as its parent in China which is called the Purple Forbidden City.

On a morning when mist lay over the Thames reaches, softening the harshness of the dock buildings and lending an air of mystery to the vessels stealing out upon the tide, a man walked briskly along Limehouse Causeway, looking about him inquiringly, as one unfamiliar with the neighborhood. Presently he seemed to recognize a turning to the right, and he pursued this for a time, now walking more slowly.

A European woman, holding a half-caste baby in her arms, stood in an open doorway, watching him uninterestedly. Otherwise, except for one neatly dressed young Chinaman, who passed him about halfway along the street, there was nothing which could have told the visitor that he had crossed the borderline dividing West from East and was now in an Oriental town.

A very narrow alleyway between two dingy houses proved to be the spot for which he was looking; and, having stared about him for a while, he entered this alleyway. At the farther end it was crossed T-fashion, by another alley, the only object of interest being an iron post at the crossing, and the scenery being made up entirely of hideous brick walls.

About halfway along on the left, set in one of these walls, were strong wooden gates, apparently those of a warehouse. Beside them was a door approached by two very dirty steps. There was a bell-push near the door, but upon neither of these entrances was there any plate to indicate the name of the proprietor of the establishment.

From his pocket-book the visitor extracted a card, consulted something written upon it, and then pressed the bell.

It was very quiet in this dingy little court. No sound of the busy thoroughfares penetrated here; and although the passage forming the top of the "T" practically marked the river bank, only dimly could one discern the sounds which belong to a seaport.

Presently the door was opened by a Chinese boy who wore the ordinary native working dress, and who regarded the man upon the step with oblique, tired-looking eyes.

"Mr. Huang Chow?" asked the caller.

The boy nodded.

"You wantchee him see?"

"If he is at home."

The boy glanced at the card, which the visitor still held between finger and thumb, and extended his hand silently. The card was surrendered. It was that of an antique dealer of Dover Street, Piccadilly, and written upon the back was the following: "Mr. Hampden would like to do business with you." The signature of the dealer followed.

The boy turned and passed along a dim and perfectly unfurnished passage which the opening of the door had revealed, while Mr. Hampden stood upon the step and lighted a cigarette.

In less than a minute the boy returned and beckoned to him to come in. As he did so, and the door was closed, he almost stumbled, so dark was the passage.

Presently, guided by the boy, he found himself in a very business-like little office, where a girl sat at an American desk, looking up at him inquiringly.

She was of a dark and arresting type. Without being pretty in the European sense, there was something appealing in her fine, dark eyes, and she possessed the inviting smile which is the heritage of Eastern women. Her dress was not unlike that of any other business girl, except that the neck of her blouse was cut very low, a fashion affected by many Eurasians, and she wore a gaily colored sash, and large and very costly pearl ear-rings. As Mr. Hampden paused in the doorway:

"Good morning," said the girl, glancing down at the card which lay upon the desk before her. "You come from Mr. Isaacs, eh?"

She looked at him with a caressing glance from beneath half- lowered lashes, but missed no detail of his appearance. She did not quite like his moustache, and thought that he would have looked better clean-shaven. Nevertheless, he was a well-set-up fellow, and her manner evidenced approval.

"Yes," he replied, smiling genially. "I have a small commission to execute, and I am told that you can help me."

The girl paused for a moment, and then:

"Yes, very likely," she said, speaking good English but with an odd intonation. "It is not jade? We have very little jade."

"No, no. I wanted an enameled casket."

"What kind?"

"Cloisonne."

"Cloisonne? Yes, we have several."

She pressed a bell, and, glancing up at the boy who had stood throughout the interview at the visitor's elbow, addressed him rapidly in Chinese. He nodded his head and led the way through a second doorway. Closing this, he opened a third and ushered Mr. Hampden into a room which nearly caused the latter to gasp with astonishment.

One who had blundered from Whitechapel into the Khan Khalil, who had been transported upon a magic carpet from a tube station to the Taj Mahal, of dropped suddenly upon Lebanon hills to find himself looking down upon the pearly domes and jeweled gardens of Damascus, could not well have been more surprised. This great treasure-house of old Huang Chow was one of Chinatown's secrets-- a secret shared only by those whose commercial interests were identical with the interests of Huang Chow.

The place was artificially lighted by lamps which themselves were beautiful objects of art, and which swung from the massive beams of the ceiling. The floor of the warehouse, which was partly of stone, was covered with thick matting, and spread upon it were rugs and carpets of Karadagh, Kermanshah, Sultan-abad, and Khorassan, with lesser-known loomings of almost equal beauty. Skins of rare beasts overlay the divans. Furniture of ivory, of ebony and lemonwood, preciously inlaid, gave to the place an air of cunning confusion. There were tall cabinets, there were caskets and chests of exquisite lacquer and enamel, loot of an emperor's palace; robes heavy with gold; slippers studded with jewels; strange carven ivories; glittering weapons; pots, jars, and bowls, as delicate and as fragile as the petals of a lily.

Last, but not least, sitting cross-legged upon a low couch, was old Huang Chow, smoking a great curved pipe, and peering half blindly across the place through large horn-rimmed spectacles. This couch was set immediately beside a wide ascending staircase, richly carpeted, and on the other side of the staircase, in a corresponding recess, upon a gilded trestle carved to represent the four claws of a dragon, rested perhaps the strangest exhibit of that strange collection--a Chinese coffin of exquisite workmanship.

The boy retired, and Mr. Hampden found himself alone with Huang Chow. No word had been exchanged between master and servant, but:

"Good morning, Mr. Hampden," said the Chinaman in a high, thin voice. "Please be seated. It is from Mr. Isaacs you come?"

CHAPTER 4

PERSONAL REPORT OF DETECTIVE JOHN DURHAM TO CHIEF INSPECTOR KERRY, OFFICER IN CHARGE OF LIMEHOUSE INQUIRY

Dear Chief Inspector,--Following your instructions I returned and interviewed the prisoner Poland in his cell. I took the line which you had suggested, pointing out to him that he had nothing to gain and everything to lose by keeping silent.

"Answer my questions," I said, "and you can walk straight out. Otherwise, you'll be up before the magistrate, and on your record alone it will mean a holiday which you probably don't want."

He was very truculent, but I got him in a good humor at last, and he admitted that he had been cooperating with the dead man, Cohen, in an attempt to burgle the house of Huang Chow. His reluctance to go into details seemed to be due rather to fear of Huang Chow than to fear of the law, and I presently gathered that he regarded Huang as responsible for the death not only of Cohen, but also of the Chinaman who was hauled out of the river about three weeks ago, as you well remember. The post-mortem showed that he had died of some kind of poisoning, and when we saw Cohen in the mortuary, his swollen appearance struck me as being very similar to that of the Chinaman. (See my report dated 31st ultimo.)

He finally agreed to talk if I would promise that he should not be charged and that his name should never be mentioned to anyone in connection with what he might tell me. I promised him that outside the ordinary official routine I would respect his request, and he told me some very curious things, which no doubt have a bearing on the case.

For instance, he had discovered--I don't know in what way--that the dead Chinaman, whose name was Pi Lung, had been in negotiation with Huang Chow for some sort of job in his warehouse. Poland had seen the man talking to Huang's daughter, at the end of the alley which leads to the place. He seemed to attach extraordinary importance to this fact. At last:

"I'll tell you what it is," he said. "That Chink was a stranger to Limehouse; I can swear to it. He was a gent of his hands; I reckon they've got 'em in China as well as here. He went out for the old boy's money-box, and finished like Cohen finished."

"Make your meaning clearer," I said.

"My meaning's this: Old Huang Chow is the biggest dealer in stolen and smuggled valuables from overseas we've got in London. He's something else as well; he's a big swell in China. But here's the point. He's got business with buyers all over London, and they have to pay cash--no checks. He doesn't bank it: I've proved that. He's got it in gold, or diamonds, or something, being wise to present conditions, hidden there in the house. Pi Lung was after his hoard. He didn't get it. Cohen and me was after it. Where's Cohen?"

I agreed that it looked very suspicious, and presently:

"When I went in with Cohen," continued Poland, "I knew one thing he didn't know--a short cut into the warehouse. He's been playing pretty-like with Lala, old Huang's daughter, and it's my belief that he knew where the store was hidden; but he never told me. We knew there were special men on duty, and we'd arranged that I was to give a signal when the patrol had passed. Cohen all the time had planned to double on me. While I was watching down on the Causeway end he climbed up and got in through the skylight I'd shown him. When I got there he was missing, but the skylight was open. I started off after him."

Then Poland clutched me, and his fright was very real.

"I heard a shriek like nothing I ever heard in my life. I saw a light shine through the trap, and then I heard a sort of moaning. Last, I heard a bang, and the light went out. I staggered down the passage half silly, started to run, and ran straight into the arms of two coppers."

This evidence I thought was conclusive, and in accordance with your instructions I proceeded to Mr. Isaacs in Dover Street. He didn't seem too pleased at my suggestion, but when I pointed out to him that one good turn deserved another, he agreed to give me an introduction to Huang Chow.

I adopted a very simple disguise, just altering my complexion and sticking on a moustache with spirit gum, hair by hair, and trimming it down military fashion. Everything ran smoothly, and I seemed to make a fairly favorable impression upon Lala Huang, the Chinaman's daughter, who evidently interviews prospective customers before they are admitted to the warehouse.

She is a Eurasian and extremely good looking. But when I found myself in the room where old Huang keeps his treasures, I really thought I was dreaming. It's a collection that must be worth thousands. He showed me snuff-bottles, cut out of gems, and with a little opening no bigger than the hole in a pipe-stem, but with wonderful paintings done inside the bottles. He'd got a model of a pagoda made out of human teeth, and a big golden rug woven from the hair of Circassian slave girls. Excuse this, Chief Inspector; I know it is what you call the romantic stuff; but I think it would have impressed you if you had seen it.

Anyway, I bought a little enameled box, in accordance with Mr. Isaacs's instructions, although whether I succeeded in convincing Huang Chow that I knew anything about the matter is more than doubtful. He got up from a sort of throne he sits on, and led the way up a broad staircase to a private room above.

"Of course, you have brought the cash, Mr. Hampden?" he said.

He speaks quite faultless English. He walked up three steps to a sort of raised writing-table in this upstairs room, and I counted out the money to him. When he sat at the table he faced toward the room, and I couldn't help thinking that, in his horn-rimmed spectacles, he looked like some old magistrate. He explained that he would pack the purchase for me, but that I must personally take it away. And:

"You understand," said he, "that you bought it from a gentleman who had purchased it abroad."

I said I quite understood. He bowed me out very politely, and presently I found myself back in the office with Lala Huang.

She seemed quite disposed to talk, and I chatted with her while the box was being packed for me to take away. I knew I must make good use of my time, but you have never given me a job I liked less. I mean, there is something very appealing about her, and I hated to think that I was playing a double game. However, without actually agreeing to see me again, she told me enough to enable me to meet her "accidentally," if I wanted to. Therefore, I am going to look out for her this evening, and probably take her to a picture palace, or somewhere where we can have a quiet talk. She seems to be fancy free, and for some reason I feel sorry for the girl. I don't altogether like the job, but I hope to justify your faith in me, Chief.

I will prepare my official report this evening when I return.

Yours obediently,--JOHN DURHAM.

CHAPTER 5

LALA HUANG

"No," said Lala Huang, "I don't like London--not this part of London."

"Where would you rather be?" asked Durham. "In China?"

Dusk had dropped its merciful curtain over Limehouse, and as the two paced slowly along West India Dock Road it seemed to the detective that a sort of glamour had crept into the scene.

He was a clever man within his limitations, and cultured up to a point; but he was not philosopher enough to know that he viewed the purlieus of Limehouse through a haze of Oriental mystery conjured up by the conversation of his companion. Temple bells there were in the clangor of the road cars. The smoke-stacks had a semblance of pagodas. Burma she had conjured up before him, and China, and the soft islands where she had first seen the light. For as well as a streak of European, there was Kanaka blood in Lala, which lent her an appeal quite new to Durham, insidious and therefore dangerous.

"Not China," she replied. "Somehow I don't think I shall ever see China again. But my father is rich, and it is dreadful to think that we live here when there are so many more beautiful places to live in."

"Then why does he stay?" asked Durham with curiosity.

"For money, always for money," answered Lala, shrugging her shoulders. "Yet if it is not to bring happiness, what good is it?"

"What good indeed?" murmured Durham.

"There is no fun for me," said the girl pathetically. "Sometimes someone nice comes to do business, but mostly they are Jews, Jews, always Jews, and..." Again she shrugged eloquently.

Durham perceived the very opening for which he had been seeking..

"You evidently don't like Jews," he said endeavoring to speak lightly.

"No," murmured the girl, "I don't think I do. Some are nice, though. I think it is the same with every kind of people--there are good and bad."

"Were you ever in America?" asked Durham.

"No."

"I was just thinking," he explained, "that I have known several American Jews who were quite good fellows."

"Yes?" said Lala, looking up at him naively, "I met one not long ago. He was not nice at all."

"Oh!" exclaimed Durham, startled by this admission, which he had not anticipated. "One of your father's customers?"

"Yes, a man named Cohen."

"Cohen?"

"A funny little chap," continued the girl. "He tried to make love to me." She lowered her lashes roguishly. "I knew all along he was pretending. He was a thief, I think. I was afraid of him."

Durham did some rapid thinking, then:

"Did you say his name was Cohen?" he asked.

"That was the name he gave."

"A man named Cohen, an American, was found dead in the river quite recently."

Lala stopped dead and clutched his arm.

"How do you know?" she demanded.

"There was a paragraph in this morning's paper."

She hesitated, then:

"Did it describe him?" she asked.

"No," replied Durham, "I don't think it did in detail. At least, the only part of the description which I remember is that he wore a large and valuable diamond on his left hand."

"Oh!" whispered Lala.

She released her grip of Durham's arm and went on.

"What?" he asked. "Did you think it was someone you knew?"

"I did know him," she replied simply. "The man who was found drowned. It is the same. I am sure now, because of the diamond ring. What paper did you read it in? I want to read it myself."

"I'm afraid I can't remember. It was probably the Daily Mail."

"Had he been drowned?"

"I presume so--yes," replied Durham guardedly.

Lala Huang was silent for some time while they paced on through the dusk. Then:

"How strange!" she said in a low voice.

"I am sorry I mentioned it," declared Durham. "But how was I to know it was your friend?"

"He was no friend of mine," returned the girl sharply. "I hated him. But it is strange nevertheless. I am sure he intended to rob my father."

"And is that why you think it strange?"

"Yes," she said, but her voice was almost inaudible.

They were come now to the narrow street communicating with the courtway in which the great treasure-house of Huang Chow was situated, and; Lala stopped at the corner.

"It was nice of you to walk along with me," she said. "Do you live in Limehouse?"

"No," replied Durham, "I don't. As a matter of fact, I came down here to-night in the hope of seeing you again."

"Did you?"

The girl glanced up at him doubtfully, and his distaste for the task set him by his superior increased with the passing of every moment. He was a man of some imagination, a great reader, and ambitious professionally. He appreciated the fact that Chief Inspector Kerry looked for great things from him, but for this type of work he had little inclination.

There was too much chivalry in his make-up to enable him to play upon a woman's sentiments, even in the interests of justice. By whatever means the man Cohen had met his death, and whether or no the Chinaman Pi Lung had died by the same hand, Lala Huang was innocent of any complicity in these matters, he was perfectly well assured.

Doubts were to come later when he was away from her, when he had had leisure to consider that she might regard him in the light of a third potential rifler of her father's treasure-house. But at the moment, looking down into her dark eyes, he reproached himself and wondered where his true duty lay.

"It is so gray and dull and sordid here," said the girl, looking down the darkened street. "There is no one much to talk to."

"But you have your business interests to keep you employed during the day, after all."

"I hate it all. I hate it all."

"But you seem to have perfect freedom?"

"Yes. My mother, you see, was not Chinese."

"But you wish to leave Limehouse?"

"I do. I do. Just now it is not so bad, but in the winter how I tire of the gray skies, the endless drizzling rain. Oh!" She shrank back into the shadow of a doorway, clutching at Durham's arm. "Don't let Ah Fu see me."

"Ah Fu? Who is Ah Fu?" asked Durham, also drawing back as a furtive figure went slinking down the opposite side of the street.

"My father's servant. He let you in this morning."

"And why must he not see you?"

"I don't trust him. I think he tells my father things."

"What is it that he carries in his hand?"

"A birdcage, I expect."

"A birdcage?"

"Yes!"

He caught the gleam of her eyes as she looked up at him out of the shadow.

"Is he, then, a bird-fancier?"

"No, no, I can't explain because I don't understand myself. But Ah Fu goes to a place in Shadwell regularly and buys young birds, always very young ones and very little ones."

"For what or for whom?"

"I don't know."

"Have you an aviary in your house?"

"No."

"Do you mean that they disappear, these purchases of Ah Fu's?"

"I often see him carrying a cage of young birds, but we have no birds in the house."

"How perfectly extraordinary!" muttered Durham.

"I distrust Ah Fu," whispered the girl. "I am glad he did not see me with you."

"Young birds," murmured Durham absently. "What kind of young birds? Any particular breed?"

"No; canaries, linnets--all sorts. Isn't it funny?" The girl laughed in a childish way. "And now I think Ah Fu will have gone in, so I must say good night."

But when presently Detective Durham found himself walking back along West India Dock Road, his mind's eye was set upon the slinking figure of a Chinaman carrying a birdcage.

CHAPTER 6

A HINT OF INCENSE

One Chinaman more or less does not make any very great difference to the authorities responsible for maintaining law and order in Limehouse. Asiatic settlers are at liberty to follow their national propensities, and to knife one another within reason. This is wisdom. Such recreations are allowed, if not encouraged, by all wise rulers of Eastern peoples.

"Found drowned," too, is a verdict which has covered many a dark mystery of old Thames, but "Found in the river, death having been due to the action of some poison unknown," is a finding which even in the case of a Chinaman is calculated to stimulate the jaded official mind.

New Scotland Yard had given Durham a roving commission, and had been justified in the fact that the second victim, and this time not a Chinaman, had been found under almost identical conditions. The link with the establishment of Huang Chow was incomplete, and Durham fully recognized that it was up to him to make it sound and incontestable.

Jim Poland was not the only man in the East End who knew that the dead Chinaman had been in negotiation with Huang Chow. Kerry knew it, and had passed the information on to Durham.

Some mystery surrounded the life of the old dealer, who was said to be a mandarin of high rank, but his exact association with the deaths first of the Chinaman Pi Lung, and second of Cohen, remained to be proved. Certain critics have declared the Metropolitan detective service to be obsolete and inefficient. Kerry, as a potential superintendent, resented these criticisms, and in his protege Durham, perceived a member of the new generation who was likely in time to produce results calculated to remove this stigma.

Durham recognized that a greater responsibility rested upon his shoulders than the actual importance of the case might have indicated; and now, proceeding warily along the deserted streets, he found his brain to be extraordinarily active and his imagination very much alive.

There is a night life in Limehouse, as he had learned, but it is a mole life, a subterranean life, of which no sign appears above ground after a certain hour. Nevertheless, as he entered the area which harbors those strange, hidden resorts the rumor of which has served to create the glamour of Chinatown, he found himself to be thinking of the great influence said to be wielded by Huang Chow, and wondering if unseen spies watched his movements.

Lala was Oriental, and now, alone in the night, distrust leapt into being within him. He had been attracted by her and had pitied her. He told himself now that this was because of her dark beauty and the essentially feminine appeal which she made. She was perhaps a vampire of the most dangerous sort, one who lured men to strange deaths for some sinister object beyond reach of a Western imagination.

He found himself doubting the success of those tactics upon which, earlier in the day, he had congratulated himself. Perhaps beneath the guise of Hampden, who bought antique furniture on commission, those cunning old eyes beneath the horn-rimmed spectacles had perceived the detective hidden, or at least had marked subterfuge.

While he could not count Lala a conquest--for he had not even attempted to make love to her--the ease with which he had developed the acquaintance now, afforded matter for suspicion.

At the entrance to the court communicating with the establishment of Huang Chow he paused, looking cautiously about him. The men on the Limehouse beats had been warned of the investigation afoot tonight, and there was a plain-clothes man on point duty at no great distance away, although carefully hidden, so that Durham had quite failed to detect his presence.

Durham wore rough clothes and rubber-soled shoes; and now, as he entered the court, he was thinking of the official report of the police sergeant who, not so many hours before, had paid a visit to the house of Huang Chow in order to question him respecting his knowledge of the dead man Cohen, and to learn when last he had seen him.

Old Huang, who had received his caller in the large room upstairs, the room which boasted the presence of the writing- dais, had exhibited no trace of confusion, assuring the sergeant that he had not seen the man Cohen for several days. Cohen had come to him with an American introduction, which he, Huang, believed to be forged, and had wanted him to undertake a shady agency, respecting the details of which he remained peculiarly reticent. In short, nothing had been gained by this official interrogation, and Huang blandly denied any knowledge of an attempted burglary of his establishment.

"What have I to lose?" he had asked the inquirer. "A lot of old lumber which I have accumulated during many years, and a reputation for being wealthy, due to my lonely habits and to the ignorance of those who live around me."

Durham, mentally reviewing the words of the report, reconstructed the scene in his mind; and now, having come to the end of the lane where the iron post rested, he stood staring up at a place in the ancient wall where several bricks had decayed, and where it was possible, according to the statement of the man Poland, to climb up on to a piece of sloping roof, and thence gain the skylight through which Cohen had obtained admittance on the night of his death.

He made sure that his automatic pistol was in his pocket, questioned the dull sounds of the riverside for a moment, looking about him anxiously, and then, using the leaning post as a stepping-stone, he succeeded in wedging his foot into a crevice in the wall. By the exercise of some agility he scrambled up to the top, and presently found himself lying upon a sloping roof.

The skylight remained well out of reach, but his rubber-soled shoes enabled him to creep up the slates until he could grasp the framework with his hands. Presently he found himself perched upon the trap which, if his information could be relied upon, possessed no fastener, or one so faulty that the trap could be raised by means of a brad-awl. He carried one in his pocket, and, screwing it into the framework, he lifted it cautiously, making very little noise.

The trap opened, and up to his nostrils there stole a queer, indefinable odor, partly that which belongs to old Oriental furniture and stuffs, but having mingled with it a hint of incense and of something else not so easily named. He recognized the smell of that strange store-room, which, as Mr. Hampden, he had recently visited.

For one moment he thought he could detect the distant note of a bell. But, listening, he heard nothing, and was reassured.

He rested the trap back against the frame, and shone the ray of an electric torch down into the darkness beneath him. The light fell upon the top of a low carven table, dragon-legged and gilded. Upon it rested the model pagoda constructed of human teeth, and there was something in this discovery which made Durham feel inclined to shudder. However, the impulse was only a passing one.

He measured the distance with his eye. The little table stood beside a deep divan, and he saw that with care it would be possible to drop upon this divan without making much noise. He calculated its exact position before replacing the torch in his pocket, and then, resting back against one side of the frame, he clutched the other with his hands. He wriggled gradually down until further purchase became impossible. He then let himself drop, and swung for a moment by his hands before releasing his hold.

He fell, as he had calculated, upon the divan. It creaked ominously. Catching his foot in the cushions, he stumbled and lay forward for a moment upon his face, listening intently.

The room was very hot but nothing stirred.

CHAPTER 7

THE SCUFFLING SOUND

Detective Durham, as he lay there inhaling the peculiar perfume of the place, recognized that he had put himself outside the pale of official protection, and was become technically a burglar.

He wondered if Chief Inspector Kerry would have approved; but he had outlined this plan of investigation for himself, and knew well that, if it were crowned by success, the end would be regarded as having justified the means. On the other hand, in the event of detention he must personally bear the consequences of such irregular behavior. He knew well, however, that his celebrated superior had achieved promotion by methods at least as irregular; and he knew that if he could but obtain evidence to account for the death of the man Cohen, and of the Chinaman Pi Lung, who had preceded him by the same mysterious path, the way of his obtaining it would not be too closely questioned.

He was an ambitious man, and consequently one who took big chances. Nothing disturbed the silence; he sat upon the divan and again pressed the button of his torch, shining it all about the low-beamed apartment and peering curiously into the weird shadows of the place. He calculated he was now in the position which Cohen had occupied during the last moments of his life, and a sense of the uncanny touched him coldly.

As he thought of the unnatural screams spoken of by Poland, some strange instinct prompted him to curl up his feet upon the divan again, as though a secret menace crawled upon the floor amid its many rugs and carpets.

He must now endeavor to reconstruct the plan upon which the American cracksman had operated. Poland had a persistent belief that Cohen had known where the fabled hoard of Huang Chow was concealed.

Durham began a deliberate inspection of the place. He thought it unlikely that a wily old Chinaman, assuming that he possessed hidden wealth, would keep it in so accessible a spot as this. It was far more probable that he had a fireproof safe in the room upstairs, perhaps built into the wall. Yet, according to Poland's account, it was in this room and not in any other that death came to Diamond Fred.

The wall-hangings first engaged Durham's attention. He moved them aside systematically, one after another, seeking for any hiding-place, but failing to find one. The door communicating with the outer office he found to be locked, but he did not believe for a moment that the office would be worthy of inspection.

There were cases containing jeweled weapons and cups and goblets inlaid with precious stones, but none of these seemed to have been tampered with, and all were locked, as was the big cabinet filled with snuff bottles.

Many of the larger pieces about the place contained drawers and cupboards, and these he systematically opened one after another, without making any discovery of note. Some of the cupboards contained broken pieces of crockery, and more or less damaged curios of one kind and another, but none of them gave him the clue for which he was seeking.

He examined the couch upon which Huang Chow had been seated when first he had met him, but although he searched it scientifically he was rewarded by no discovery.

A very fusty and unpleasant smell was more noticeable at this point than elsewhere in the room, and he found himself staring speculatively up the wide, carpeted stairs. Next he turned his attention to the lacquered coffin which occupied the corresponding recess to that filled by the couch. It was an extraordinarily ornate piece of lacquer work and probably of great value.

The lid appeared to be screwed on, and Durham stood staring at the thing, half revolted and half fascinated. He failed to discover any means of opening it, however, and when he tried to move it bodily found it very heavy. He came to the conclusion that all the portable valuables were contained in locked cases or cabinets, and out of this discovery grew an idea.

The case containing the snuff bottles stood too close to the wall to enable him t test his new theory, but a square case near the office door, in which were five of six small but almost priceless pieces of porcelain, afforded the very evidence for which he was looking.

Thin electric flex descended from somewhere inside the case down one of the legs of the pedestal, and through a neatly drilled hole in the floor, evidently placed there to accommodate it.

"Burglar alarm!" he muttered.

The opening of this case, and doubtless of any of the others, would set alarm bells ringing. This was not an unimportant discovery, but it brought him very little nearer to a solution of the chief problem which engaged his mind. Assuming that Cohen had opened one of the cases and had alarmed old Huang Chow, what steps had the latter taken to deal with the intruder which had resulted in so ghastly a death? And how had he disposed of the body?

As Durham stood there musing and looking down through the plate- glass at the delicate porcelain beneath, a faint sound intruded itself upon the stillness. It gave him another idea. Part of the floor was stone-paved, but part was wood.

Upon a portion of the latter, where no carpet rested, Durham dropped flat, pressing his ear to the floor.

A faint swishing and trickling sound was perceptible from some place beneath.

"Ah!" he murmured.

Remembering that the premises almost overhung the Thames, he divined that the cellars were flooded at high tide, or that there was some kind of drain or cutting running underneath the house.

He stood up again, listening intently for any sound within the building. He thought he had detected something, and now, as he stood there alert, he heard it again--a faint scuffling, which might have been occasioned by rats or even mice, but which, in some subtle and very unpleasant way, did not suggest the movements of these familiar rodents.

Even as he perceived it, it ceased, leaving him wondering, and uncomfortably conscious of a sudden dread of his surroundings. He wondered in what part of this mysterious house Lala resided, and recognizing that his departure must leave traces, he determined to prosecute his inquiries as far as possible, since another opportunity might not arise.

He was baffled but still hopeful. Something there was in the smell of the place which threatened to unnerve him; or perhaps in its silence, which remained quite unbroken save when, by acute listening, one detected the dripping of water.

That unexplained scuffling sound, too, which he had failed to trace or identify, lingered in his memory insistently, and for some reason contained the elements of fear.

He crossed the room and began softly to mount the stair. It creaked only slightly, and the door at the top proved to be ajar. He peeped in, to find the place empty. It was a typical Chinese apartment, containing very little furniture, the raised desk being the most noticeable item, except for a small shrine which faced it on the other side of the room.

He mounted the steps to the desk and inspected a number of loose papers which lay upon it. Without exception they were written in Chinese. A sort of large, dull white blotting-pad lay upon the table, but its surface was smooth and glossy.

Over it was suspended what looked like a lampshade, but on inspection it proved to contain no lamp, but to communicate, by a sort of funnel, with the ceiling above.

At this contrivance Durham stared long and curiously, but without coming to any conclusion respecting its purpose. He might have investigated further, but he became aware of a dull and regular sound in the room behind him.

He turned in a flash, staring in the direction of two curtains draped before what he supposed to be a door.

On tiptoe he crossed and gently drew the curtains aside.

He looked into a small, cell-like room, lighted by one window, where upon a low bed Huang Chow lay sleeping peacefully!

Durham almost held his breath; then, withdrawing as quietly as he had approached, he descended the stair. At the foot his attention was again arrested by the faint scuffling sound. It ceased as suddenly as it had begun, leaving him wondering and conscious anew of a chill of apprehension.

He had already made his plans for departure, but knew that they must leave evidence, when discovered, of his visit.

A large and solid table stood near the divan, and he moved this immediately under the trap. Upon it he laid a leopard-skin to deaden any noise he might make, and then upon the leopard-skin he set a massive chair: he replaced his torch in his pocket and drew himself up on to the roof again. Reclosing the trap by means of the awl which he had screwed into it, he removed the awl and placed it in his pocket.

Then, sliding gently down the sloping roof, he dropped back into the deserted court.

CHAPTER 8

A CAGE OF BIRDS

"No," said Lala, "we have never had robbers in the house." She looked up at Durham naively. "You are not a thief, are you?" she asked.

"No, I assure you I am not," he answered, and felt himself flushing to the roots of his hair.

They were seated in a teashop patronized by the workers of the district; and as Durham, his elbows resting on the marble-topped table, looked into the dark eyes of his companion, he told himself again that whatever might be the secrets of old Huang Chow, his daughter did not share them.

The Chinaman had made no report to the authorities, although the piled up furniture beneath the skylight must have afforded conclusive evidence that a burglarious entry had been made into the premises.

"I should feel very nervous," Durham declared, "with all those valuables in the house."

"I feel nervous about my father," the girl answered in a low voice. "His room opens out of the warehouse, but mine is shut away in another part of the building. And Ah Fu sleeps behind the office."

"Were you not afraid when you suspected that Cohen was a burglar? You told me yourself that you did suspect him."

"Yes, I spoke to my father about it."

"And what did he say?"

"Oh"--she shrugged her shoulders--"he just smiled and told me not to worry."

"And that was the last you heard about the matter?"

"Yes, until you told me he was dead."

Again he questioned the dark eyes and again was baffled. He felt tempted, and not for the first time, to throw up the case. After all, it rested upon very slender data--the mysterious death of a Chinaman whose history was unknown and the story of a crook whose word was worth nothing.

Finally he asked himself, as he had asked himself before, what did it matter? If old Huang Chow had disposed of these people in some strange manner, they had sought to rob him. The morality of the case was complicated and obscure, and more and more he was falling under the spell of Lala's dark eyes.

But always it was his professional pride which came to the rescue. Murder had been done, whether justifiably or otherwise, and to him had been entrusted the discovery of the murderer. It seemed that failure was to be his lot, for if Lala knew anything she was a most consummate actress, and if she did not, his last hope of information was gone.

He would have liked nothing better than to be rid of the affair, provided he could throw up the case with a clear conscience. But when presently he parted from the attractive Eurasian, and watched her slim figure as, turning, she waved her hand and disappeared round a corner, he knew that rest was not for him.

He had discovered the emporium of a Shadwell live-stock dealer with whom Ah Fu had a standing order for newly fledged birds of all descriptions. Purchases apparently were always made after dusk, and Ah Fu with his birdcage was due that evening.

A scheme having suggested itself to Durham, he now proceeded to put it into execution, so that when dusk came, and Ah Fu, carrying an empty birdcage, set out from the house of Huang Chow, a very dirty-looking loafer passed the corner of the street at about the time that the Chinaman came slinking out.

Durham had mentally calculated that Ah Fu would be gone about half an hour upon his mysterious errand, but the Chinaman traveled faster than he had calculated.

Just as he was about to climb up once more on to the sloping roof, he heard the pattering footsteps returning to the courtyard, although rather less than twenty minutes had elapsed since the man had set out.

Durham darted round the corner and waited until he heard the door closed; then, returning, he scrambled up on to the roof, creeping forward until he was lying looking down through the skylight into the darkened room below.

For ten minutes or more he waited, until he began to feel cramped and uncomfortable. Then that happened which he had hoped and anticipated would happen. The place beneath became illuminated, not fully, by means of the hanging lamps, but dimly so that distorted shadows were cast about the floor. Someone had entered carrying a lantern.

Durham's view-point limited his area of vision, but presently, as the light came nearer and nearer, he discerned Ah Fu, carrying a lantern in one hand and a birdcage in the other. He could hear nothing, for the trap fitted well and the glass was thick. Moreover, it was very dirty. He was afraid, however, to attempt to clean a space.

Ah Fu apparently had set the lantern upon a table, and into the radius of its light there presently moved a stooping figure. Durham recognized Huang Chow, and felt his heart beats increasing in rapidity.

Clutching the framework of the trap with his hands, he moved his head cautiously, so that presently he was enabled to see the two Chinamen. They were standing beside the lacquered coffin upon its dragon-legged pedestal. Durham stifled an exclamation.

One end of the ornate sarcophagus had been opened in some way!

Now, to the watcher's unbounded astonishment, Ah Fu placed the birdcage in the opening, and apparently reclosed the trap in the end of the coffin. He made other manipulations with his bony yellow fingers, which Durham failed to comprehend. Finally the birdcage was withdrawn again, and as it was passed before the light of the lantern he saw that it was empty, whereas previously it had contained a number of tiny birds all huddled up together!

The light gleamed upon the spectacles of Huang Chow. Watching him, Durham saw him take out from a hidden drawer in the pedestal a long, slender key, insert it in a lock concealed by the ornate carving, and then slightly raise the lid which had so recently defied his own efforts.

He raised it only a few inches, and then, taking up the lantern, peered into the interior of the coffin, at the same time waving his hand in dismissal to Ah Fu. For a while he stood there, peering into the interior, and then, lowering the lid again, he relocked this gruesome receptacle and, lantern in hand, began to mount the stair.

Durham inhaled deeply. He realized that during the last few seconds he had been holding his breath. Now, as he began to creep back down the slope, he discovered that his hands were shaking.

He dropped down into the court again, and for several minutes leaned against the wall, endeavoring to reason out an explanation of what he had seen, and in a measure to regain his composure.

There was a horror underlying it all which he was half afraid to face. But the real clue to the mystery still eluded him.

Whether what he had witnessed were some kind of obscene ceremony, or whether an explanation more vile must be sought, he remained undetermined. He must repeat his exploit, if possible, and once more gain access to the room which contained the lacquer coffin.

But the adventure was very distasteful. He recollected the smell of the place, and the memory brought with it a sense of nausea. He thought of Lala Huang, and his ideas became grotesque and chaotic. Yet the solution of the mystery lay at last within his grasp, and to the zest of the investigator everything else became subjugated.

He walked slowly away, silent in his rubber-soled shoes.

CHAPTER 9

THE PICTURE ON THE PAD

Lala Huang lay listening to the vague sounds which disturbed the silence of the night. Presently her thoughts made her sigh wearily. During the lifetime of her mother, who had died while Lala was yet a little girl, life had been different and so much brighter.

She imagined that in the mingled sounds of dock and river which came to her she could hear the roar of surf upon a golden beach. The stuffy air of Limehouse took on the hot fragrance of a tropic island, and she sighed again, but this time rapturously, for in spirit she was a child once more, lulled by the voice of the great Pacific.

Young as she was, the death of her mother had been a blow from which it had taken her several years to recover. Then had commenced those long travels with her father, from port to port, from ocean to ocean, sometimes settling awhile, but ever moving onward, onward.

He had had her educated after a fashion, and his love for her she did not doubt. But her mother's blood spoke more strongly than that part of her which was Chinese, and there was softness and a delicious languor in her nature which her father did not seem to understand, and of which he did not appear to approve.

She knew that he was wealthy. She knew that his ways were not straight ways, although that part of his business to which he had admitted her as an assistant, and an able one, was legitimate enough, or so it seemed.

Consignments of goods arrived at strange hours of the night at the establishment in Limehouse, and from this side of her father's transactions she was barred. The big double doors opening on the little courtyard would be opened by Ah Fu, and packing cases of varying sizes be taken in. Sometimes the sounds of these activities would reach her in her room in a distant part of the house; but only in the morning would she recognize their significance, when in the warehouse she would discover that some new and choice pieces had arrived.

She wondered with what object her father accumulated wealth, and hoped, against the promptings of her common sense, that he designed to return East, there to seek a retirement amidst the familiar and the beautiful things of the Orient which belonged to Lala's dream of heaven.

Stories about her father often reached her ears. She knew that he had held high rank in China before she had been born; but that he had sacrificed his rights in some way had always been her theory. She had been too young to understand the stories which her mother had told her sometimes; but that there were traits in the character of Huang Chow which it was not good for his daughter to know she appreciated and accepted as a secret sorrow.

He allowed her all the freedom to which her education entitled her. Her life was that of a European and not of an Oriental woman. She loved him in a way, but also feared him. She feared the dark and cruel side of his character, of which, at various periods during their life together, she had had terrifying glimpses.

She had decided that cruelty was his vice. In what way he gratified it she had never learned, nor did she desire to do so. There were periodical visits from the police, but she had learned long ago that her father was too clever to place himself within reach of the law.

However crooked one part of his business methods might be, his dealings with his clients were straight enough, so that no one had any object in betraying him; and the legality or otherwise of his foreign relations evidently afforded no case against him upon which the authorities could act, or upon which they cared to act.

In America it had been graft which had protected him. She had learned this accidentally, but never knew whether he bought his immunity in the same way in London.

Some of the rumors which reached her were terrifying. Latterly she had met many strange glances in her comings and goings about Limehouse. This peculiar atmosphere had always preceded the break-up of every home which they had shared. She divined the fact that in some way Huang Chow had outstayed his welcome in Chinatown, London. Where their next resting-place would be she could not imagine, but she prayed that it might be in some more sunny clime.

She found herself to be thinking over much of John Hampden. His bona fides were not above suspicion, but she could scarcely expect to meet a really white man in such an environment.

Lala would have liked to think that he was white, but could not force herself to do so. She would have liked to think that he sought her company because she appealed to him personally; but she had detected the fact that another motive underlay his attentions. She wondered if he could be another of those moths drawn by the light of that fabled wealth of her father.

It was curious, she reflected, that Huang Chow never checked-- indeed, openly countenanced--her friendship with the many chance acquaintances she had made, even when her own instincts told her that the men were crooked; so that, knowing the acumen of her father, she was well aware that he must know it too.

Several of these pseudo lovers of hers had died. It was a point which often occurred to her mind, but upon which she did not care to dwell even now. But John Hampden--John Hampden was different. He was not wholly sincere. She sighed wearily. But nevertheless he was not like some of the others.

She started up in bed, seized with a sudden dreadful idea. He was a detective!

She understood now why she had found so much that was white in him, but so much that was false. His presence seemed to be very near her. Something caressing in his voice echoed in her mind. She found herself to be listening to the muted sounds of Limehouse and of the waterway which flowed so close beside her.

That old longing for the home of her childhood returned tenfold, and tears began to trickle down her cheeks. She was falling in love with this man whose object was her father's ruin. A cold terror clutched at her heart. Even now, while their friendship was so new, so strange, there was a query, a stark, terrifying query, to stand up before her.

If put to the test, which would she choose?

She was unable to face that issue, and dropped back upon her pillow, stifling a sob.

Yes, he was a detective. In some way her father had at last attracted the serious attention of the law. Rumors of this were flying round Chinatown, to which she had not been entirely deaf. She thought of a hundred questions, a hundred silences, and grew more and more convinced of the truth.

What did he mean to do? Before her a ghostly company uprose--the shadows of some she had known with designs upon her father. John Hampden's design was different. But might he not join that mysterious company?

Now again she suddenly sprang upright, this time because of a definite sound which had reached her ears from within the house: a very faint, bell-like tinkling which ceased almost immediately. She had heard it one night before, and quite recently; indeed, on the night before she had met John Hampden. Cohen--Cohen, the Jew, had died that night!

She sprang lightly on to the floor, found her slippers, and threw a silk kimono over her nightrobe. She tiptoed cautiously to the door and opened it.

It was at this very moment that old Huang Chow, asleep in his cell-like apartment, was aroused by the tinkling of a bell set immediately above his head. He awoke instantly, raised his hand and stopped the bell. His expression, could anyone have been present to see it, was a thing unpleasant to behold. Triumph was in it, and cunning cruelty.

His long yellow fingers reached out for his horn-rimmed spectacles which lay upon a little table beside him. Adjusting them, he pulled the curtains aside and shuffled silently across the large room.

Mounting the steps to the raised writing-table, he rested his elbows upon it, and peered down at that curious blotting-pad which had so provoked the curiosity of Durham. Could Durham have seen it now the mystery must have been solved. It was an ingenious camera obscura apparatus, and dimly depicted upon its surface appeared a reproduction of part of the storehouse beneath! The part of it which was visible was that touched by the light of an electric torch, carried by a man crossing the floor in the direction of the lacquered coffin upon the gilded pedestal!

Old Huang Chow chuckled silently, and his yellow fingers clutched the table edge as he moved to peer more closely into the picture.

"Poor fool!" he whispered in Chinese. "Poor fool!"

It was the man who had come with the introduction from Mr. Isaacs--a new impostor who sought to rob him, who sought to obtain information from his daughter, who had examined his premises last night, and had even penetrated upstairs, so that he, old Huang Chow, had been compelled to disconnect the apparatus and to feign sleep under the scrutiny of the intruder.

To-night it would be otherwise. To-night it would be otherwise.

CHAPTER 10

THE LACQUERED COFFIN

Durham gently raised the trap in the roof of Huang Chow's treasure-house. He was prepared for snares and pitfalls. No sane man, on the evidence which he, Durham, had been compelled to leave behind, would have neglected to fasten the skylight which so obviously afforded a means of entrance into his premises.

Therefore, he was expected to return. The devilish mechanism was set ready to receive him. But the artist within him demanded that he should unmask the mystery with his own hands.

Moreover, he doubted that an official visit, even now, would yield any results. Old Huang Chow was too cunning for that. If he was to learn how the man Cohen had died, he must follow the same path to the bitter end. But there were men on duty round the house, and he believed that he had placed them so secretly as to deceive even this master of cunning with whom he was dealing.

He repeated his exploit, dropping with a dull thud upon the cushioned divan. Then, having lain there listening awhile, he pressed the button of his torch, and, standing up, crept across the room in the direction of the stairway.

Here he paused awhile, listening intently. The image of Lala Huang arose before his mind's eye reproachfully, but he crushed the reproach, and advanced until he stood beside the lacquered coffin.

He remembered where the key was hidden, and, stooping, he fumbled for a while and then found it. He was acutely conscious of an unnameable fear. He felt that he was watched, and yet was unwilling to believe it. The musty and unpleasant smell which he had noticed before became extremely perceptible.

He quietly sought for the hidden lock, and, presently finding it, inserted the key, then paused awhile. He rested his torch upon the cushions of the divan where the light shone directly upon the coffin. Then, having his automatic in his left hand, he turned the key.

He had expected now to be able to raise the lid as he had seen Huang Chow do; but the result was far more surprising.

The lid, together with a second framework of fine netting, flew open with a resounding bang; and from the interior of the coffin uprose a most abominable stench.

Durham started back a step, and as he did so witnessed a sight which turned him sick with horror.

Out on to the edge of the coffin leapt the most gigantic spider which he had ever seen in his life! It had a body as big as a man's fist, jet black, with hairy legs like the legs of a crab and a span of a foot or more!

A moment it poised there, while he swayed, sick with horror. Then, unhesitatingly, it leapt for his face!

He groaned and fired, missed the horror, but diverted its leap, so that it fell with a sickening thud a yard behind him. He turned, staggering back towards the stair, and aware that a light had shone out from somewhere.

A door had been opened only a few yards from where he stood, and there, framed in the opening, was Lala Huang, her eyes wide with terror and her gaze set upon him across the room.

"You!" she whispered. "You!"

"Go back!" he cried hoarsely. "Go back! Close the door. You don't understand--close the door!"

Her gaze set wildly upon him, Lala staggered forward; stopped dead; looked down at her bare ankle, and then, seeing the thing which had fastened upon her, uttered a piercing shriek which rang throughout the place.

At which moment the floor slid away beneath Durham, and he found himself falling-- falling--and then battling for life in evil- smelling water, amidst absolute darkness.

Police whistles were skirling around the house of Huang Chow. As the hidden men came running into the court:

"You heard the shot?" cried the sergeant in charge. "I warned him not to go alone. Don't waste time on the door. One man stay on duty there; the rest of you follow me."

In a few moments, led by the sergeant, the party came dropping heavily through the skylight into the treasure-house of Huang Chow, in which every lamp was now alight. A trap was open near the foot of the stairs, and from beneath it muffled cries proceeded. In this direction the sergeant headed. Craning over the trap:

"Hallo, Mr. Durham!" he called. "Mr. Durham!"

"Get a rope and a ladder," came a faint cry from below. "I can just touch bottom with my feet and keep my head above water, but the tide's coming in. Look to the girl, though, first. Look to the girl!"

The sergeant turned to where, stretched upon a tiger skin before a half-open door, Lala Huang lay, scantily clothed and white as death.

Upon one of her bare ankles was a discolored mark.

As the sergeant and another of the men stooped over her a moaning sound drew their attention to the stair, and there, bent and tottering slowly down, was old Huang Chow, his eyes peering through the owl-like glasses vacantly across the room to where his daughter lay.

"My God!" whispered the sergeant, upon one knee beside her. He looked blankly into the face of the other man. "She's dead!"

Two plain-clothes men were busy knotting together tapestries and pieces of rare stuff with which to draw Durham out of the pit; but at these old Huang Chow looked not at all, but gropingly crossed the room, as if he saw imperfectly, or could not believe what he saw. At last he reached the side of the dead girl, stooped, touched her, laid a trembling yellow hand over her heart, and then stood up again, looking from face to face.

Ignoring the mingled activities about him, he crossed to the open coffin and began to fumble amongst the putrefying mass of bones and webbing which lay therein. Out from this he presently drew an iron coffer.

Carrying it across the room he opened the lid. It was full almost to the top with uncut gems of every variety--diamonds, rubies, sapphires, emeralds, topaz, amethysts, flashing greenly, redly, whitely. In handfuls he grasped them and sprinkled them upon the body of the dead girl.

"For you," he crooned brokenly in Chinese. "They were all for you!"

The extemporized rope had just been lowered to Durham, when:

"My God!" cried the sergeant, looking over Huang Chow's shoulder. "What's that?"

He had seen the giant spider, the horror from Surinam, which the Chinaman had reared and fed to guard his treasure and to gratify his lust for the strange and cruel. The insect, like everything else in that house, was unusual, almost unique. It was one of the Black Soldier spiders, by some regarded as a native myth, but actually existing in Surinam and parts of Brazil. A member of the family, Mygale, its sting was more quickly and certainly fatal than that of a rattle-snake. Its instinct was fearlessly to attack any creature, great or small, which disturbed it in its dark hiding-place.

Now, with feverish, horrible rapidity it was racing up the tapestries on the other side of the room.

"Merciful God!" groaned the sergeant.

Snatching a revolver from his pocket he fired shot after shot. The third hit the thing but did not kill it. It dropped back upon the floor and began to crawl toward the coffin. The sergeant ran across and at close quarters shot it again.

Red blood oozed out from the hideous black body and began to form a deep stain upon the carpet.

When Durham, drenched but unhurt, was hauled back into the treasure-house, he did not speak, but, scrambling into the room stood--pallid--staring dully at old Huang Chow.

Huang Chow, upon his knees beside his daughter, was engaged in sprinkling priceless jewels over her still body, and murmuring in Chinese:

"For you, for you, Lala. They were all for you."

KERRY'S KID

CHAPTER I

RED KERRY ON DUTY

Chief Inspector Kerry came down from the top of a motor-bus and stood on the sidewalk for a while gazing to right and left along Piccadilly. The night was, humid and misty, now threatening fog and now rain. Many travelers were abroad at this Christmas season, the pleasure seekers easily to be distinguished from those whom business had detained in town, and who hurried toward their various firesides. The theatres were disgorging their audiences. Streams of lighted cars bore parties supperward; less pretentious taxicabs formed links in the chain.

From the little huddled crowd of more economical theatre-goers who waited at the stopping place of the motor-buses, Kerry detached himself, walking slowly along westward and staring reflectively about him. Opposite the corner of Bond Street he stood still, swinging his malacca cane and gazing fixedly along this narrow bazaar street of the Baghdad of the West. His trim, athletic figure was muffled in a big, double-breasted, woolly overcoat, the collar turned up about his ears. His neat bowler hat was tilted forward so as to shade the fierce blue eyes. Indeed, in that imperfect light, little of the Chief Inspector's countenance was visible except his large, gleaming white teeth, which he constantly revealed in the act of industriously chewing mint gum.

He smiled as he chewed. Duty had called him out into the midst, and for once he had obeyed reluctantly. That very afternoon had seen the return of Dan Kerry, junior, home from school for the Christmas vacation, and Dan was the apple of his father's eye.

Mrs. Kerry had reserved her dour Scottish comments upon the boy's school report for a more seemly occasion than the first day of his holidays; but Kerry had made no attempt to conceal his jubilation--almost immoral, his wife had declared it to be-- respecting the lad's athletic record. His work on the junior left wing had gained the commendation of a celebrated international; and Kerry, who had interviewed the gymnasium instructor, had learned that Dan Junior bade fair to become an amateur boxer of distinction.

"He is faster on his feet than any boy I ever handled," the expert had declared. "He hasn't got the weight behind it yet, of course, but he's developing a left that's going to make history. I'm of opinion that there isn't a boy in the seniors can take him on, and I'll say that he's a credit to you."

Those words had fallen more sweetly upon the ears of Chief Inspector Kerry than any encomium of the boy's learning could have done. On the purely scholastic side his report was not a good one, admittedly. "But," murmured Kerry aloud, "he's going to be a man."

He remembered that he had promised, despite the lateness of the hour, to telephone the lad directly he had received a certain report, and to tell him whether he might wait up for his return or whether he must turn in. Kerry, stamping his small, neatly shod feet upon the pavement, smiled agreeably. He was thinking of the telephone which recently he had had installed in his house in Brixton. His wife had demanded this as a Christmas box, pointing out how many uneasy hours she would be spared by the installation. Kerry had consented cheerfully enough, for was he not shortly to be promoted to the exalted post of a superintendent of the Criminal Investigation Department?

These reflections were cheering and warming; and, waiting until a gap occurred in the stream of cabs and cars, he crossed Piccadilly and proceeded along Bond Street, swinging his shoulders in a manner which would have enabled any constable in the force to recognize "Red Kerry" at a hundred yards.

The fierce eyes scrutinized the occupants of all the lighted cars. At pedestrians also he stared curiously, and at another smaller group of travelers waiting for the buses on the left- hand side of the street he looked hard and long. He pursued his way, acknowledged the salutation of a porter who stood outside the entrance to the Embassy Club, and proceeded, glancing about him right and left and with some evident and definite purpose.

A constable standing at the corner of Conduit Street touched his helmet as Kerry passed and the light of an arc-lamp revealed the fierce red face. The Chief Inspector stopped, turned, and:

"What the devil's the idea?" he demanded.

He snapped out the words in such fashion that the unfortunate constable almost believed he could see sparks in the misty air.

"I'm sorry, sir, but recognizing you suddenly like, I----"

"You did?" the fierce voice interrupted. "How long in the force?"

"Six months, sir."

"Never salute an officer in plain clothes."

"I know, sir."

"Then why did you do it?"

"I told you, sir."

"Then tell me again."

"I forgot."

"You're paid to remember; bear it in mind."

Kerry tucked his malacca under his arm and walked on, leaving the unfortunate policeman literally stupefied by his first encounter with the celebrated Chief Inspector.

Presently another line of cars proclaimed the entrance to a club, and just before reaching the first of these Kerry paused. A man stood in a shadowy doorway, and:

"Good evening, Chief Inspector," he said quietly.

"Good evening, Durham. Anything to report?"

"Yes. Lou Chada is here again.

"With whom?"

"Lady Rourke."

Kerry stepped to the edge of the pavement and spat out a piece of chewing-gum. From his overcoat pocket he drew a fresh piece, tore off the pink wrapping and placed the gum between his teeth. Then:

"How long?" he demanded.

"Came to dinner. They are dancing."

"H'm!" The Chief Inspector ranged himself beside the other detective in the shadow of the doorway. "Something's brewing, Durham," he said. "I think I shall wait."

His subordinate stared curiously but made no reply. He was not wholly in his chief's confidence. He merely knew that the name of Lou Chada to Kerry was like a red rag to a bull. The handsome, cultured young Eurasian, fresh from a distinguished university career and pampered by a certain section of smart society, did not conform to Detective Sergeant Durham's idea of a suspect. He knew that Lou was the son of Zani Chada, and he knew that Zani Chada was one of the wealthiest men in Limehouse. But Lou had an expensive flat in George Street; Lou was courted by society butterflies, and in what way he could be connected with the case known as "the Limehouse inquiry," Durham could not imagine.

That the open indiscretion of Lady "Pat" Rourke might lead to trouble with her husband, was conceivable enough; but this was rather a matter for underhand private inquiry than for the attention of the Criminal Investigation Department of New Scotland Yard.

So mused Durham, standing cold and uncomfortable in the shadowy doorway, and dreaming of a certain cozy fireside, a pair of carpet slippers and a glass of hot toddy which awaited him. Suddenly:

"Great flames! Look!" he cried.

Kerry's fingers closed, steely, upon Durham's wrist. A porter was urgently moving the parked cars farther along the street to enable one, a French coupe, to draw up before the club entrance.

Two men came out, supporting between them a woman who seemed to be ill; a slender, blonde woman whose pretty face was pale and whose wide-open blue eyes stared strangely straight before her. The taller of her escorts, while continuing to support her, solicitously wrapped her fur cloak about her bare shoulders; the other, the manager of the club, stepped forward and opened the door of the car.

"Lady Rourke!" whispered Durham.

"With Lou Chada!" rapped Kerry. "Run for a cab. Brisk. Don't waste a second."

Some little conversation ensued between manager and patron, then the tall, handsome Eurasian, waving his hand protestingly, removed his hat and stepped into the coupe beside Lady Rourke. It immediately moved away in the direction of Piccadilly.

One glimpse Kerry had of the pretty, fair head lying limply back against the cushions. The manager of the club was staring after the car.

Kerry stepped out from his hiding place. Durham had disappeared, and there was no cab in sight, but immediately beyond the illuminated entrance stood a Rolls-Royce which had been fifth in the rank of parked cars before the adjustment had been made to enable the coupe to reach the door. Kerry ran across, and:

"Whose car, my lad?" he demanded of the chauffeur.

The latter, resenting the curt tone of the inquiry, looked the speaker up and down, and:

"Captain. Egerton's," he replied slowly. "But what business may it be of yours?"

"I'm Chief Inspector Kerry, of New Scotland Yard," came the rapid reply. "I want to follow the car that has just left."

"What about running?" demanded the man insolently.

Kerry shot out a small, muscular hand and grasped the speaker's wrist.

"I'll say one thing to you," he rapped. "I'm a police officer, and I demand your help. Refuse it, and you'll wake up in Vine Street."

The Chief Inspector was on the step now, bending forward so that his fierce red face was but an inch removed from that of the startled chauffeur. The quelling force of his ferocious personality achieved its purpose, as it rarely failed to do.

"I'm getting in," added the Chief Inspector, jumping back on to the pavement. "Lose that French bus, and I'll charge you with resisting and obstructing an officer of the law in the execution of his duty. Start."

Kerry leaped in and banged the door--and the Rolls-Royce started.

CHAPTER 2

AT MALAY JACK'S

When Kerry left Bond Street the mistiness of the night was developing into definite fog. It varied in different districts. Thus, St. Paul's Churchyard had been clear of it at a time when it had lain impenetrably in Trafalgar Square. When, an hour and a half after setting out in the commandeered Rolls-Royce, Kerry groped blindly along Limehouse Causeway, it was through a yellow murk that he made his way--a vapor which could not only be seen, smelled and felt, but tasted.

He was in one of his most violent humors. He found some slight solace in the reflection that the impudent chauffeur, from whom he had parted in West India Dock Road, must experience great difficulty in finding his way back to the West End.

"Damn the fog!" he muttered, coughing irritably.

It had tricked him, this floating murk of London; for, while he had been enabled to keep the coupe in view right to the fringe of dockland, here, as if bred by old London's river, the fog had lain impenetrably.

Chief Inspector Kerry was a man who took many risks, but because of this cursed fog he had no definite evidence that Chada's car had gone to a certain house. Right of search he had not, and so temporarily he was baffled.

Now the nearest telephone was his objective, and presently, where a blue light dimly pierced the mist, he paused, pushed open a swing door, and stepped into a long, narrow passage. He descended three stairs, and entered a room laden with a sickly perfume compounded of stale beer and spirits; of greasy humanity--European, Asiatic, and African; of cheap tobacco and cheaper scents; and, vaguely, of opium.

It was fairly well lighted, but the fog had penetrated here, veiling some of the harshness of its rough appointments. An unsavory den was Malay Jack's, where flotsam of the river might be found. Yellow men there were, and black men and brown men. But all the women present were white.

Fan-tan was in progress at one of the tables, the four players being apparently the only strictly sober people in the room; A woman was laughing raucously as Kerry entered, and many coarse- voiced conversations were in progress; but as he pulled the rough curtain walls aside and walked into the room, a hush, highly complimentary to the Chief Inspector's reputation, fell upon the assembly. Only the woman's raucous laughter continued, rising, a hideous solo, above a sort of murmur, composed of the words "Red Kerry!" spoken in many tones.

Kerry ignored the sensation which his entrance had created, and crossed the room to a small counter, behind which a dusky man was standing, coatless and shirt sleeves rolled up. He had the skin of a Malay but the features of a stage Irishman of the old school. And, indeed, had he known his own pedigree, which is a knowledge beyond the ken of any man, partly Irish he might have found himself indeed to be.

This was Malay Jack, the proprietor of one of the roughest houses in Limehouse. His expression, while propitiatory, was not friendly, but:

"Don't get hot and bothered," snapped Kerry viciously. "I want to use your telephone, that's all."

"Oh," said the other, unable to conceal his relief, "that's easy. Come in."

He raised a flap in the counter, and Kerry, passing through, entered a little room behind the bar. Here a telephone stood upon a dirty, littered table, and, taking it up:

"City four hundred," called the Chief Inspector curtly. A moment later: "Hallo! Yes," he said. "Chief Inspector Kerry speaking. Put me through to my department, please."

He stood for a while waiting, receiver in hand, and smiled grimly to note that the uproar in the room beyond had been resumed. Evidently Malay Jack had given the "all clear" signal. Then:

"Chief Inspector Kerry speaking," he said again. "Has Detective Sergeant Durham reported?"

"Yes," was the reply, "half an hour ago. He's standing-by at Limehouse Station. He followed you in a taxi, but lost you on the way owing to the fog."

"I don't wonder," said Kerry. "His loss is not so great as mine. Anything else?"

"Nothing else."

"Good. I'll speak to Limehouse. Good-bye."

He replaced the receiver and paused for a moment, reflecting. Extracting a piece of tasteless gum from between his teeth, he deposited it in the grate, where a sickly fire burned; then, tearing the wrapper from a fresh slip, he resumed his chewing and stood looking about him with unseeing eyes. Fierce they were as ever, but introspective in expression.

Famous for his swift decisions, for once in a way he found himself in doubt. Malay Jack had keen ears, and there were those in the place who had every reason to be interested in the movements of a member of the Criminal Investigation Department, especially of one who had earned the right to be dreaded by the rats of Limehouse. London's peculiar climate fought against him, but he determined to make no more telephone calls but to proceed to Limehouse police station.

He stepped swiftly into the bar, and, as he had anticipated, nearly upset the proprietor, who was standing listening by the half-open door. Kerry smiled fiercely into the ugly face, lifted the flap, and walked down the room, through the aisle between the scattered tables, where the air was heavy with strange perfumes, touched now with the bite of London fog, and where slanting eyes and straight eyes, sober eyes and drunken eyes, regarded him furtively. Something of a second hush there was, but one not so complete as the first.

Kerry pulled the curtain aside, mounted the stair, walked along the passage and out through the swing door into the yellow gloom of the Causeway. Ten slow steps he had taken when he detected a sound of pursuit. Like a flash he turned, clenching his fists. Then:

"Inspector!" whispered a husky voice.

"Yes! Who are you? What do you want?"

A dim form loomed up through the fog.

"My name is Peters, sir. Inspector Preston knows me."

Kerry had paused immediately under a street lamp, and now he looked into the pinched, lean face of the speaker, and:

"I've heard of you," he snapped. "Got some information for me?"

"I think so; but walk on."

Chief Inspector Kerry hesitated. Peters belonged to a class which Kerry despised with all the force of his straightforward character. A professional informer has his uses from the police point of view; and while evidence of this kind often figured in reports made to the Chief Inspector, he personally avoided contact with such persons, as he instinctively and daintily avoided contact with personal dirt. But now, something so big was at stake that his hesitation was only momentary.

A vision of the pale face of Lady Rourke, of the golden head leaning weakly back upon the cushions of the coupe, as he had glimpsed it in Bond Street, rose before his mind's eye as if conjured up out of the fog. Peters shuffled along beside him, and:

"Young Chada's done himself in to-night," continued the husky voice. "He brought a swell girl to the old mans house an hour ago. I was hanging about there, thinking I might get some information. I think she was doped."

"Why?" snapped Kerry.

"Well, I was standing over on the other side of the street. Lou Chada opened the door with a key; and when the light shone out I saw him carry her in."

"Carry her in?"

"Yes. She was in evening dress, with a swell cloak."

"The car?"

"He came out again and drove it around to the garage at the back."

"Why didn't you report this at once?"

"I was on my way to do it when I saw you coming out of Malay Jack's."

The man's voice shook nervously, and:

"What are you scared about?" asked Kerry savagely. "Got anything else to tell me?"

"No, no," muttered Peters. "Only I've got an idea he saw me."

"Who saw you?"

"Lou Chada."

"What then?"

"Well, only--don't leave me till we get to the station."

Kerry blew down his nose contemptuously, then stopped suddenly.

"Stand still," he ordered. "I want to listen."

Silent, they stood in a place of darkness, untouched by any lamplight. Not a sound reached them through the curtain of fog. Asiatic mystery wrapped them about, but Kerry experienced only contempt for the cowardice of his companion, and:

"You need come no farther," he said coldly. "Good night."

"But..." began the man.

"Good night," repeated Kerry.

He walked on briskly, tapping the pavement with his malacca. The sneaking figure of the informer was swallowed up in the fog. But not a dozen paces had the Chief Inspector gone when he was arrested by a frenzied scream, rising, hollowly, in a dreadful, muffled crescendo. Words reached him.

"My God, he's stabbed me!"

Then came a sort of babbling, which died into a moan.

"Hell!" muttered Kerry, "the poor devil was right!"

He turned and began to run back, fumbling in his pocket for his electric torch. Almost in the same moment that he found it he stumbled upon Peters, who lay half in the road and half upon the sidewalk.

Kerry pressed the button, and met the glance of upturned, glazing eyes. Even as he dropped upon his knee beside the dying man, Peters swept his arm around in a convulsive movement, having the fingers crooked, coughed horribly, and rolled upon his face.

Switching off the light of the torch, Kerry clenched his jaws in a tense effort of listening, literally holding his breath. But no sound reached him through the muffling fog. A moment he hesitated, well knowing his danger, then viciously snapping on the light again, he quested in the blood-stained mud all about the body of the murdered man.

"Ah!"

It was an exclamation of triumph.

One corner hideously stained, for it had lain half under Peters's shoulder, Kerry gingerly lifted between finger and thumb a handkerchief of fine white silk, such as is carried in the breast pocket of an evening coat.

It bore an ornate monogram worked in gold, and representing the letters "L. C." Oddly enough, it was the corner that bore the monogram which was also bloodstained.

CHAPTER 3

THE ROOM OF THE GOLDEN BUDDHA

It was a moot point whether Lady Pat Rourke merited condemnation or pity. She possessed that type of blonde beauty which seems to be a lodestone for mankind in general. Her husband was wealthy, twelve years her senior, and, far from watching over her with jealous care--an attitude which often characterizes such unions-- he, on the contrary, permitted her a dangerous freedom, believing that she would appreciate without abusing it.

Her friendship with Lou Chada had first opened his eyes to the perils which beset the road of least resistance. Sir Noel Rourke was an Anglo-Indian, and his prejudice against the Eurasian was one not lightly to be surmounted. Not all the polish which English culture had given to this child of a mixed union could blind Sir Noel to the yellow streak. Courted though Chada was by some of the best people, Sir Noel remained cold.

The long, magnetic eyes, the handsome, clear-cut features, above all, that slow and alluring smile, appealed to the husband of the willful Pat rather as evidences of Oriental, half-effeminate devilry than as passports to decent society. Oxford had veneered him, but scratch the veneer and one found the sandal-wood of the East, perfumed, seductive, appealing, but something to be shunned as brittle and untrustworthy.

Yet he hesitated, seeking to be true to his convictions. Knowing what he knew already, and what he suspected, it is certain that, could he have viewed Lou Chada through the eyes of Chief Inspector Kerry, the affair must have terminated otherwise. But Sir Noel did not know what Kerry knew. And the pleasure-seeking Lady Rourke, with her hair of spun gold and her provoking smile, found Lou Chada dangerously fascinating; almost she was infatuated--she who had known so much admiration.

Of those joys for which thousands of her plainer sisters yearn and starve to the end of their days she had experienced a surfeit. Always she sought for novelty, for new adventures. She was confident of herself, but yet--and here lay the delicious thrill--not wholly confident. Many times she had promised to visit the house of Lou Chada's father--a mystery palace cunningly painted, a perfumed page from the Arabian poets dropped amid the interesting squalor of Limehouse.

Perhaps she had never intended to go. Who knows? But on the night when she came within the ken of Chief Inspector Kerry, Lou Chada had urged her to do so in his poetically passionate fashion, and, wanting to go, she had asked herself: "Am I strong enough? Dare I?"

They had dined, danced, and she had smoked one of the scented cigarettes which he alone seemed to be able to procure, and which, on their arrival from the East, were contained in queer little polished wooden boxes.

Then had come an unfamiliar nausea and dizziness, an uncomfortable recognition of the fact that she was making a fool of herself, and finally a semi-darkness through which familiar faces loomed up and were quickly lost again. There was the soft, musical voice of Lou Chada reassuring her, a sense of chill, of helplessness, and then for a while an interval which afterward she found herself unable to bridge.

Knowledge of verity came at last, and Lady Pat raised herself from the divan upon which she had been lying, and, her slender hands clutching the cushions, stared about her with eyes which ever grew wider.

She was in a long, rather lofty room, which was lighted by three silver lanterns swung from the ceiling. The place, without containing much furniture, was a riot of garish, barbaric color. There were deep divans cushioned in amber and blood-red. Upon the floor lay Persian carpets and skins of beasts. Cunning niches there were, half concealing and half revealing long-necked Chinese jars; and odd little carven tables bore strangely fashioned vessels of silver. There was a cabinet of ebony inlaid with jade, there were black tapestries figured with dragons of green and gold. Curtains she saw of peacock-blue; and in a tall, narrow recess, dominating the room, squatted a great golden Buddha.

The atmosphere was laden with a strange perfume.

But, above all, this room was silent, most oppressively silent.

Lady Pat started to her feet. The whole perfumed place seemed to be swimming around her. Reclosing her eyes, she fought down her weakness. The truth, the truth respecting Lou Chada and herself, had uprisen starkly before her. By her own folly--and she could find no tiny excuse--she had placed herself in the power of a man whom, instinctively, deep within her soul, she had always known to be utterly unscrupulous.

How cleverly he had concealed the wild animal which dwelt beneath that suave, polished exterior! Yet how ill he had concealed it! For intuitively she had always recognized its presence, but had deliberately closed her eyes, finding a joy in the secret knowledge of danger. Now at last he had discarded pretense.

The cigarette which he had offered her at the club had been drugged. She was in Limehouse, at the mercy of a man in whose veins ran the blood of ancestors to whom women had been chattels. Too well she recognized that his passion must have driven him insane, as he must know at what cost he took such liberties with one who could not lightly be so treated. But these reflections afforded poor consolation. It was not of the penalties that Lou Chada must suffer for this infringement of Western codes, but of the price that she must pay for her folly, of which Pat was thinking.

There was a nauseating taste upon her palate. She remembered having noticed it faintly while she was smoking the cigarette; indeed, she had commented upon it at the time.

"The dirty yellow blackguard!" she said aloud, and clenched her hands.

She merely echoed what many a man had said before her. She wondered at herself, and in doing so but wondered at the mystery of womanhood.

Clarity was returning. The room no longer swam around her. She crossed in the direction of a garish curtain, which instinctively she divined to mask a door. Dragging it aside, she tried the handle, but the door was locked. A second door she found, and this also proved to be locked.

There was one tall window, also covered by ornate draperies, but it was shuttered, and the shutters had locks. Another small window she discovered, glazed with amber glass, but set so high in the wall as to be inaccessible.

Dread assailed her, and dropping on to one of the divans, she hid her face in her hands.

"My God!" she whispered. "My God! Give me strength--give me courage."

For a long time she remained there, listening for any sound which should disperse the silence. She thought of her husband, of the sweet security of her home, of the things which she had forfeited because of this mad quest of adventure. And presently a key grated in a lock.

Lady Pat started to her feet with a wild, swift action which must have reminded a beholder of a startled gazelle. The drapery masking the door which she had first investigated was drawn aside. A man entered and dropped the curtain behind him.

Exactly what she had expected she could not have defined, but the presence of this perfect stranger was a complete surprise. The man, who wore embroidered slippers and a sort of long blue robe, stood there regarding her with an expression which, even in her frantic condition, she found to be puzzling. He had long, untidy gray hair brushed back from his low brow; eyes strangely like the eyes of Lou Chada, except that they were more heavy-lidded; but his skin was as yellow as a guinea, and his gaunt, clean-shaven face was the face of an Oriental.

The slender hands, too, which he held clasped before him, were yellow, and possessed a curiously arresting quality. Pat imagined them clasped about her white throat, and her very soul seemed to shrink from the man who stood there looking at her with those long, magnetic, inscrutable eyes.

She wondered why she was surprised, and suddenly realized that it was because of the expression in his eyes, for it was an expression of cold anger. Then the intruder spoke.

"Who are you?" he demanded, speaking with an accent which was unfamiliar to her, but in a voice which was not unlike the voice of Lou Chada. "Who brought you here?"

This was so wholly unexpected that for a moment she found herself unable to reply, but finally:

"How dare you!" she cried, her native courage reasserting itself. "I have been drugged and brought to this place. You shall pay for it. How dare you!"

"Ah!" The long, dark eyes regarded her unmovingly. "But who are you?"

"I am Lady Rourke. Open the door. You shall bitterly regret this outrage."

"You are Lady Rourke?" the man repeated. "Before you speak of regrets, answer the question which I have asked: Who brought you here?"

"Lou Chada."

"Ah!" There was no alteration of pose, no change of expression, but slightly the intonation had varied.

"I don't know who you are, but I demand to be released from this place instantly."

The man standing before the curtained door slightly inclined his head.

"You shall be released," he replied, "but not instantly. I will see the one who brought you here. He may not be entirely to blame. Before you leave we shall understand one another."

Tone and glance were coldly angry. Then, before the frightened woman could say another word, the man in the blue robe robe withdrew, the curtain was dropped again, and she heard the grating of a key in the lock. She ran to the door, beating upon it with her clenched hands.

"Let me go!" she cried, half hysterically. "Let me go! You shall pay for this! Oh, you shall pay for this!"

No one answered, and, turning, she leaned back against the curtain, breathing heavily and fighting for composure, for strength.

CHAPTER 4

ZANI CHADA, THE EURASIAN

I can't help thinking, Chief Inspector," said the officer in charge at Limehouse Station, "that

44

you take unnecessary risks."

"Can't you?" said Kerry, tilting his bowler farther forward and staring truculently at the speaker.

"No, I can't. Since you cleaned up the dope gang down here you've been a marked man. These murders in the Chinatown area, of which this one to-night makes the third, have got some kind of big influence behind them. Yet you wander about in the fog without even a gun in your pocket."

"I don't believe in guns," rapped Kerry. "My bare hands are good enough for any yellow smart in this area. And if they give out I can kick like a mule."

The other laughed, shaking his head.

"It's silly, all the same," he persisted. "The man who did the job out there in the fog to-night might have knifed you or shot you long before you could have got here."

"He might," snapped Kerry, "but he didn't."

Yet, remembering his wife, who would be waiting for him in the cozy sitting-room he knew a sudden pang. Perhaps he did take unnecessary chances. Others had said so. Hard upon the thought came the memory of his boy, and of the telephone message which the episodes of the night had prevented him from sending.

He remembered, too, something which his fearless nature had prompted him to forget: he remembered how, just as he had arisen from beside the body of the murdered man, oblique eyes had regarded him swiftly out of the fog. He had lashed out with a boxer's instinct, but his knuckles had encountered nothing but empty air. No sound had come to tell him that the thing had not been an illusion. Only, once again, as he groped his way through the shuttered streets of Chinatown and the silence of the yellow mist, something had prompted him to turn; and again he had detected the glint of oblique eyes, and faintly had discerned the form of one who followed him.

Kerry chewed viciously, then:

"I think I'll 'phone the wife," he said abruptly. "She'll be expecting me."

Almost before he had finished speaking the 'phone bell rang, and a few moments later:

"Someone to speak to you, Chief Inspector," cried the officer in charge.

"Ah!" exclaimed Kerry, his fierce eyes lighting up. "That will be from home."

"I don't think so," was the reply. "But see who it is."

"Hello!" he called.

He was answered by an unfamiliar voice, a voice which had a queer, guttural intonation. It was the sort of voice he had learned to loathe.

"Is that Chief Inspector Kerry?"

"Yes," he snapped.

"May I take it that what I have to say will be treated in confidence?"

45

"Certainly not."

"Think again, Chief Inspector," the voice continued. "You are a man within sight of the ambition of years, and although you may be unaware of the fact, you stand upon the edge of a disaster. I appreciate your sense of duty and respect it. But there are times when diplomacy is a more potent weapon than force."

Kerry, listening, became aware that the speaker was a man of cultured intellect. He wondered greatly, but:

"My time is valuable," he said rapidly. "Come to the point. What do you want and who are you?"

"One moment, Chief Inspector. An opportunity to make your fortune without interfering with your career has come in your way. You have obtained possession of what you believe to be a clue to a murder."

The voice ceased, and Kerry remaining silent, immediately continued:

"Knowing your personal character, I doubt if you have communicated the fact of your possessing this evidence to anyone else. I suggest, in your own interests, that before doing so you interview me."

Kerry thought rapidly, and then:

"I don't say you're right," he rapped back. "But if I come to see you, I shall leave a sealed statement in possession of the officer in charge here."

"To this I have no objection," the guttural voice replied, "but I beg of you to bring the evidence with you."

"I'm not to be bought," warned Kerry. "Don't think it and don't suggest it, or when I get to you I'll break you in half."

His red moustache positively bristled, and he clutched the receiver so tightly that it quivered against his ear.

"You mistake me," replied the speaker. "My name is Zani Chada. You know where I live. I shall not detain you more than five minutes if you will do me the honor of calling upon me."

Kerry chewed furiously for ten momentous seconds, then:

"I'll come!" he said.

He replaced the receiver on the hook, and, walking across to the charge desk, took an official form and a pen. On the back of the form he scribbled rapidly, watched with curiosity by the officer in charge.

"Give me an envelope," he directed.

An envelope was found and handed to him. He placed the paper in the envelope, gummed down the lapel, and addressed it in large, bold writing to the Assistant Commissioner of the Criminal Investigation Department, who was his chief. Finally:

"I'm going out," he explained.

"After what I've said?"

"After what you've said. I'm going out. If I don't come back or don't telephone within the next hour, you will know what to do with this."

The Limehouse official stared perplexedly.

"But meanwhile," he protested, "what steps am I to take about the murder? Durham will be back with the body at any moment now, and you say you've got a clue to the murderer."

"I have," said Kerry, "but I'm going to get definite evidence. Do nothing until you hear from me."

"Very good," answered the other, and Kerry, tucking his malacca cane under his arm, strode out into the fog.

His knowledge of the Limehouse area was extensive and peculiar, so that twenty minutes later, having made only one mistake in the darkness, he was pressing an electric bell set beside a door which alone broke the expanse of a long and dreary brick wall, lining a street which neither by day nor night would have seemed inviting to the casual visitor.

The door was opened by a Chinaman wearing national dress, revealing a small, square lobby, warmly lighted and furnished Orientally. Kerry stepped in briskly.

"I want to see Mr. Zani Chada. Tell him I am here. Chief Inspector Kerry is my name."

The Chinaman bowed, crossed the lobby, and, drawing some curtains aside, walked up four carpeted stairs and disappeared into a short passage revealed by the raising of the tapestry. As he did so Kerry stared about him curiously.

He had never before entered the mystery house of Zani Chada, nor had he personally encountered the Eurasian, reputed to be a millionaire, but who chose, for some obscure reason, to make his abode in this old rambling building, once a country mansion, which to-day was closely invested by dockland and the narrow alleys of Chinatown. It was curiously still in the lobby, and, as he determined, curiously Eastern. He was conscious of a sense of exhilaration. That Zani Chada controlled powerful influences, he knew well. But, reviewing the precautions which he had taken, Kerry determined that the trump card was in his possession.

The Chinese servant descended the stairs again and intimated that the visitor should follow him. Kerry, carrying his hat and cane, mounted the stairs, walked along the carpeted passage, and was ushered into a queer, low room furnished as a library.

It was lined with shelves containing strange-looking books, none of which appeared to be English. Upon the top of the shelves were grotesque figures of gods, pieces of Chinese pottery and other Oriental ornaments. Arms there were in the room, and rich carpets, carven furniture, and an air of luxury peculiarly exotic. Furthermore, he detected a faint smell of opium from which fact he divined that Zani Chada was addicted to the national vice of China.

Seated before a long narrow table was the notorious Eurasian. The table contained a number of strange and unfamiliar objects, as well as a small rack of books. An opium pipe rested in a porcelain bowl.

Zani Chada, wearing a blue robe, sat in a cushioned chair, staring toward the Chief Inspector. With one slender yellow hand he brushed his untidy gray hair. His long magnetic eyes were half closed.

"Good evening, Chief Inspector Kerry," he said. "Won't you be seated?"

"Thanks, I'm not staying. I can hear what you've got to say standing."

The long eyes grew a little more narrow--the only change of expression that Zani Chada allowed himself.

"As you wish. I have no occasion to detain you long."

In that queer, perfumed room, with the suggestion of something sinister underlying its exotic luxury, arose a kind of astral clash as the powerful personality of the Eurasian came in contact with that of Kerry. In a sense it was a contest of rapier and battle-axe; an insidious but powerful will enlisted against the bulldog force of the Chief Inspector.

Still through half-closed eyes Zani Chada watched his visitor, who stood, feet apart and chin thrust forward aggressively, staring with wide open, fierce blue eyes at the other.

"I'm going to say one thing," declared Kerry, snapping out the words in a manner little short of ferocious. He laid his hat and cane upon a chair and took a step in the direction of the narrow, laden table. "Make me any kind of offer to buy back the evidence you think I've got, and I'll bash your face as flat as a frying- pan."

The yellow hands of Zani Chada clutched the metal knobs which ornamented the arms of the chair in which he was seated. The long eyes now presented the appearance of being entirely closed; otherwise he remained immovable.

Following a short, portentous silence:

"How grossly you misunderstood me, Chief Inspector," Chada replied, speaking very softly. "You are shortly to be promoted to a post which no one is better fitted to occupy. You enjoy great domestic happiness, and you possess a son in whom you repose great hopes. In this respect Chief Inspector, I resemble you."

Kerry's nostrils were widely dilated, but he did not speak.

"You see," continued the Eurasian, "I know many things about you. Indeed, I have watched your career with interest. Now, to be brief, a great scandal may be averted and a woman's reputation preserved if you and I, as men of the world, can succeed in understanding one another."

"I don't want to understand you," said Kerry bluntly. "But you've said enough already to justify me in blowing this whistle." He drew a police whistle from his overcoat pocket. "This house is being watched."

"I am aware of the fact," murmured Zani Chada.

"There are two people in it I want for two different reasons. If you say much more there may be three."

Chada raised his hand slowly.

"Put back your whistle, Chief Inspector."

There was a curious restraint in the Eurasian's manner which Kerry distrusted, but for which at the time he was at a loss to account. Then suddenly he determined that the man was waiting for something, listening for some sound. As if to confirm this reasoning, just at that moment a sound indeed broke the silence of the room.

Somewhere far away in the distance of the big house a gong was beaten three times softly. Kerry's fierce glance searched the face of Zani Chada, but it remained mask-like, immovable. Yet that this had been a signal of some kind the Chief Inspector did not doubt, and:

"You can't trick me," he said fiercely. "No one can leave this house without my knowledge, and because of what happened out there in the fog my hands are untied."

He took up his hat and cane from the chair.

"I'm going to search the premises," he declared.

Zani Chada stood up slowly.

"Chief Inspector," he said, "I advise you to do nothing until you have consulted your wife."

"Consulted my wife?" snapped Kerry. "What the devil do you mean?"

"I mean that any steps you may take now can only lead to disaster for many, and in your own case to great sorrow."

Kerry took a step forward, two steps, then paused. He was considering certain words which the Eurasian had spoken. Without fearing the man in the physical sense, he was not fool enough to underestimate his potentialities for evil and his power to strike darkly.

"Act as you please," added Zani Chada, speaking even more softly. "But I have not advised lightly. I will receive you, Chief Inspector, at any hour of the night you care to return. By to- morrow, if you wish, you may be independent of everybody."

Kerry clenched his fists.

"And great sorrow may be spared to others," concluded the Eurasian.

Kerry's teeth snapped together audibly; then, putting on his hat, he turned and walked straight to the door.

CHAPTER 5

DAN KERRY, JUNIOR

Dan Kerry, junior, was humorously like his father, except that he was larger-boned and promised to grow into a much bigger man. His hair was uncompromisingly red, and grew in such irregular fashion that the comb was not made which could subdue it. He had the wide-open, fighting blue eyes of the Chief Inspector, and when he smiled the presence of two broken teeth lent him a very pugilistic appearance.

On his advent at the school of which he was now one of the most popular members, he had promptly been christened "Carrots." To this nickname young Kerry had always taken exception, and he proceeded to display his prejudice on the first day of his arrival with such force and determination that the sobriquet had been withdrawn by tacit consent of every member of the form who hitherto had favored it.

"I'll take you all on," the new arrival had declared amidst a silence of stupefaction, "starting with you"--pointing to the biggest boy. "If we don't finish to-day, I'll begin again to-morrow."

The sheer impudence of the thing had astounded everybody. Young Kerry's treatment of his leading persecutor had produced a salutary change of opinion. Of such kidney was Daniel Kerry, junior; and when, some hours after his father's departure on the night of the murder in the fog, the 'phone bell rang, it was Dan junior, and not his mother, who answered the call.

"Hallo!" said a voice. "Is that Chief Inspector Kerry's house?"

"Yes," replied Dan.

"It has begun to rain in town," the voice continued, "Is that the Chief Inspector's son speaking?"

"Yes, I'm Daniel Kerry."

"Well, my boy, you know the way to New Scotland Yard?"

"Rather."

"He says will you bring his overall? Do you know where to find it?"

"Yes, yes!" cried Dan excitedly, delighted to be thus made a party to his father's activities.

"Well, get it. Jump on a tram at the Town Hall and bring the overall along here. Your mother will not object, will she?"

"Of course not," cried Dan. "I'll tell her. Am I to start now?"

"Yes, right away."

Mrs. Kerry was sewing by the fire in the dining room when her son came in with the news, his blue eyes sparkling excitedly. She nodded her head slowly.

"Ye'll want ye'r Burberry and ye'r thick boots," she declared, "a muffler, too, and ye'r oldest cap. I think it's madness for ye to go out on such a night, but~~~~"

"Father said I could," protested the boy.

"He says so, and ye shall go, but I think it madness a' the same."

However, some ten minutes later young Kerry set out, keenly resenting the woolen muffler which he had been compelled to wear, and secretly determined to remove it before mounting the tram. Across one arm he carried the glistening overall which was the Chief Inspector's constant companion on wet nights abroad. The fog had turned denser, and ten paces from the door of the house took him out of sight of the light streaming from the hallway.

Mary Kerry well knew her husband's theories about coddling boys, but even so could not entirely reconcile herself to the present expedition. However, closing the door, she returned philosophically to her sewing, reflecting that little harm could come to Dan after all, for he was strong, healthy, and intelligent.

On went the boy through the mist, whistling merrily. Not twenty yards from the house a coupe was drawn up, and by the light of one of its lamps a man was consulting a piece of paper on which, presumably, an address was written; for, as the boy approached, the man turned, his collar pulled up about his face, his hat pulled down.

"Hallo!" he called. "Can you please tell me something?"

He spoke with a curious accent, unfamiliar to the boy. "A foreigner of some kind," young Kerry determined.

"What is it?" he asked, pausing.

"Will you please read and tell me if I am near this place?" the man continued, holding up the paper which he had been scrutinizing.

Dan stepped forward and bent over it. He could not make out the writing, and bent yet more, holding it nearer to the lamp. At which moment some second person neatly pinioned him from behind, a scarf was whipped about his head, and, kicking furiously but otherwise helpless, he felt himself lifted and placed inside the car.

The muffler had been thrown in such fashion about his face as to leave one eye partly free, and as he was lifted he had a momentary glimpse of his captors. With a thrill of real, sickly terror he realized that he was in the hands of Chinamen!

Perhaps telepathically this spasm of fear was conveyed to his father, for it was at about this time that the latter was interviewing Zani Chada, and at about this time that Kerry recognized, underlying the other's words, at once an ill- concealed suspense and a threat. Then, a few minutes later, had come the three strokes of the gong; and again that unreasonable dread had assailed him, perhaps because it signalized the capture of his son, news of which had been immediately telephoned to Limehouse by Zani Chada's orders.

Certain it is that Kerry left the Eurasian's house in a frame of mind which was not familiar to him. He was undecided respecting his next move. A deadly menace underlay Chada's words.

"Consult your wife," he kept muttering to himself. When the door was opened for him by the Chinese servant, he paused a moment before going out into the fog. There were men on duty at the back and at the front of the house. Should he risk all and raid the place? That Lady Rourke was captive here he no longer doubted. But it was equally certain that no further harm would come to her at the hands of her captors, since she had been traced there and since Zani Chada was well aware of the fact. Of the whereabouts of Lou Chada he could not be certain. If he was in the house, they had him.

The door was closed by the Chinaman, and Kerry stood out in the darkness of the dismal, brick-walled street, feeling something as nearly akin to dejection as was possible in one of his mercurial spirit. Something trickled upon the brim of his hat, and, raising his head,

Kerry detected rain upon his upturned face. He breathed a prayer of thankfulness. This would put an end to the fog.

He began to walk along by the high brick wall, but had not proceeded far before a muffled figure arose before him and the light of an electric torch was shone into his face.

"Oh, it's you, Chief Inspector!" came the voice of the watcher.

"It is," rapped Kerry. "Unless there are tunnels under this old rat-hole, I take it the men on duty can cover all the exits?"

"All the main exits," was the reply. "But, as you say, it's a strange house, and Zani Chada has a stranger reputation."

"Do nothing until you hear from me."

"Very good, Chief Inspector."

The rain now was definitely conquering the fog, and in half the time which had been occupied by the outward journey Kerry was back again in Limehouse. police station. Unconsciously he had been hastening his pace with every stride, urged onward by an unaccountable anxiety, so that finally he almost ran into the office and up to the desk where the telephone stood.

Lifting it, he called his own number and stood tapping his foot, impatiently awaiting the reply. Presently came the voice of the operator: "Have they answered yet?"

"No."

"I will ring them again."

Kerry's anxiety became acute, almost unendurable; and when at last, after repeated attempts, no reply could be obtained from his home, he replaced the receiver and leaned for a moment on the desk, shaken with such a storm of apprehension as he had rarely known. He turned to the inspector in charge, and:

"Let me have that envelope I left with you," he directed. "And have someone 'phone for a taxi; they are to keep on till they get one. Where is Sergeant Durham?"

"At the mortuary."

"Ah!"

"Any developments, Chief Inspector?"

"Yes. But apart from keeping a close watch upon the house of Zani Chada you are to do nothing until you hear from me again."

"Very good," said the inspector. "Are you going to wait for Durham's report?"

"No. Directly the cab arrives I am going to wait for nothing."

Indeed, he paced up and down the room like a wild beast caged, while call after call was sent to neighboring cab ranks, for a long time without result. What did it mean, his wife's failure to answer the telephone? It might mean that neither she nor their one servant nor Dan was in the house. And if they were not in the house at this hour of the night, where could they possibly be? This it might mean, or--something worse.

A thousand and one possibilities, hideous, fantastic, appalling, flashed through his mind. He was beginning to learn what Zani Chada had meant when he had said: "I have followed your career with interest."

At last a taxi was found, and the man instructed over the 'phone to proceed immediately to Limehouse station. He seemed so long in coming that when at last the cab was heard to pause outside, Kerry could not trust himself to speak to the driver, but directed a sergeant to give him the address. He entered silently and closed the door.

A steady drizzle of rain was falling. It had already dispersed the fog, so that he might hope with luck to be home within the hour. As a matter of fact, the man performed the journey in excellent time, but it seemed to his passenger that he could have walked quicker, such was the gnawing anxiety within him and the fear which prompted him to long for wings.

Instructing the cabman to wait, Kerry unlocked the front door and entered. He had noted a light in the dining room window, and entering, he found his wife awaiting him there. She rose as he entered, with horror in her comely face.

"Dan!" she whispered. "Dan! where is ye'r mackintosh?"

"I didn't take it," he replied, endeavoring to tell himself that his apprehensions had been groundless. "But how was it that you did not answer the telephone?"

"What do ye mean, Dan?" Mary Kerry stared, her eyes growing wider and wider. "The boy answered, Dan. He set out wi' ye'r mackintosh full an hour and a half since."

"What!"

The truth leaped out at Kerry like an enemy out of ambush.

"Who sent that message?"

"Someone frae the Yard, to tell the boy to bring ye'r mackintosh alone at once. Dan! Dan..."

She advanced, hands outstretched, quivering, but Kerry had leaped out into the narrow hallway. He raised the telephone receiver, listened for a moment, and then jerked it back upon the hook.

"Dead line!" he muttered. "Someone has been at work with a wire- cutter outside the house!"

His wife came out to where he stood, and, clenching his teeth very grimly, he took her in his arms. She was shaking as if palsied.

"Mary dear," he said, "pray with all your might that I am given strength to do my duty."

She looked at him with haggard, tearless eyes.

"Tell me the truth: ha' they got my boy?"

His fingers tightened on her shoulders.

"Don't worry," he said, "and don't ask me to stay to explain. When I come back I'll have Dan with me!"

He trusted himself no further, but, clapping his hat on his head, walked out to the waiting cab.

"Back to Limehouse police station," he directed rapidly.

"Lor lumme!" muttered the taximan. "Where are you goin' to after that, guv'nor? It's a bit off the map."

"I'm going to hell!" rapped Kerry, suddenly thrusting his red face very near to that of the speaker. "And you're going to drive me!"

CHAPTER 6

THE KNIGHT ERRANT

Recognizing the superior strength of his captors, young Kerry soon gave up struggling. The thrill of his first real adventure entered into his blood. He remembered that he was the son of his father, and he realized, being a quick-witted lad, that he was in the grip of enemies of his father. The panic which had threatened him when first he had recognized that he was in the hands of Chinese, gave place to a cold rage--a heritage which in later years was to make him a dangerous man.

He lay quite passively in the grasp of someone who held him fast, and learned, by breathing quietly, that the presence of the muffler about his nose and mouth did not greatly inconvenience him. There was some desultory conversation between the two men in the car, but it was carried on in an odd, sibilant language which the boy did not understand, but which he divined to be Chinese. He thought how every other boy in the school would envy him, and the thought was stimulating, nerving. On the very first day of his holidays he was become the central figure of a Chinatown drama.

The last traces of fear fled. His position was uncomfortable and his limbs were cramped, but he resigned himself, with something almost like gladness, and began to look forward to that which lay ahead with a zest and a will to be no passive instrument which might have surprised his captors could they have read the mind of their captive.

The journey seemed almost interminable, but young Kerry suffered it in stoical silence until the car stopped and he was lifted and carried down stone steps into some damp, earthy-smelling place. Some distance was traversed, and then many flights of stairs were mounted, some bare but others carpeted.

Finally he was deposited in a chair, and as he raised his hand to the scarf, which toward the end of the journey had been bound more tightly about his head so as to prevent him from seeing at all, he heard a door closed and locked.

The scarf was quickly removed. And Dan found himself in a low- ceilinged attic having a sloping roof and one shuttered window. A shadeless electric lamp hung from the ceiling. Excepting the cane-seated chair in which he had been deposited and a certain amount of nondescript lumber, the attic was unfurnished. Dan rapidly considered what his father would have done in the circumstances.

"Make sure that the door is locked," he muttered.

54

He tried it, and it was locked beyond any shadow of doubt.

"The window."

Shutters covered it, and these were fastened with a padlock.

He considered this padlock attentively; then, drawing from his pocket one of those wonderful knives which are really miniature tool-chests, he raised from a grove the screw-driver which formed part of its equipment, and with neatness and dispatch unscrewed the staple to which the padlock was attached!

A moment later he had opened the shutters and was looking out into the drizzle of the night.

The room in which he was confined was on the third floor of a dingy, brick-built house; a portion of some other building faced him; down below was a stone-paved courtyard. To the left stood a high wall, and beyond it he obtained a glimpse of other dingy buildings. One lighted window was visible--a square window in the opposite building, from which amber light shone out.

Somewhere in the street beyond was a standard lamp. He could detect the halo which it cast into the misty rain. The glass was very dirty, and young Kerry raised the sash, admitting a draught of damp, cold air into the room. He craned out, looking about him eagerly.

A rainwater-pipe was within reach of his hand on the right of the window and, leaning out still farther, young Kerry saw that it passed beside two other, larger, windows on the floor beneath him. Neither of these showed any light.

Dizzy heights have no terror for healthy youth. The brackets supporting the rain-pipe were a sufficient staircase for the agile Dan, a more slippery prisoner than the famous Baron Trenck; and, discarding his muffler and his Burberry, he climbed out upon the sill and felt with his thick-soled boots for the first of these footholds. Clutching the ledge, he lowered himself and felt for the next.

Then came the moment when he must trust all his weight to the pipe. Clenching his teeth, he risked it, felt for and found the third angle, and then, still clutching the pipe, stood for a moment upon the ledge of the window immediately beneath him. He was curious respecting the lighted window of the neighboring house; and, twisting about, he bent, peering across--and saw a sight which arrested his progress.

The room within was furnished in a way which made him gasp with astonishment. It was like an Eastern picture, he thought. Her golden hair disheveled and her hands alternately clenching and unclenching, a woman whom he considered to be most wonderfully dressed was pacing wildly up and down, a look of such horror upon her pale face that Dan's heart seemed to stop beating for a moment!

Here was real trouble of a sort which appealed to all the chivalry in the boy's nature. He considered the window, which was glazed with amber-colored glass, observed that it was sufficiently open to enable him to slip the fastening and open it entirely could he but reach it. And--yes!--there was a rain- pipe!

Climbing down to the yard, he looked quickly about him, ran across, and climbed up to the lighted window. A moment later he had pushed it widely open.

He was greeted by a stifled cry, but, cautiously transferring his weight from the friendly pipe to the ledge, he got astride of it, one foot in the room. Then, by exercise of a monkey-like agility, he wriggled his head and shoulders within.

"It's all right," he said softly and reassuringly; "I'm Dan Kerry, son of Chief Inspector Kerry. Can I be of any assistance?"

Her hands clasped convulsively together, the woman stood looking up at him.

"Oh, thank God!" said the captive. "But what are you going to do? Can you get me out?"

"Don't worry," replied Dan confidently. "Father and I can manage it all right!"

He performed a singular contortion, as a result of which his other leg and foot appeared inside the window. Then, twisting around, he lowered himself and dropped triumphantly upon a cushioned divan. At that moment he would have faced a cage full of man-eating tigers. The spirit of adventure had him in its grip. He stood up, breathing rapidly, his crop of red hair more disheveled than usual.

Then, before he could stir or utter any protest, the golden- haired princess whom he had come to rescue stooped, threw her arms around his neck, and kissed him.

"You darling, brave boy!" she said. "I think you have saved me from madness."

Young Kerry, more flushed than ever, extricated himself, and:

"You're not out of the mess yet," he protested. "The only difference is that I'm in it with you!"

"But where is your father?"

"I'm looking for him."

"What!"

"Oh! he's about somewhere," Dan assured her confidently.

"But, but~~~~" She was gazing at him wide-eyed, "Didn't he send you here?"

"You bet he didn't," returned young Kerry. "I came here on my own accord, and when I go you're coming with me. I can't make out how you got here, anyway. Do you know whose house this is?"

"Oh, I do, I do!"

"Whose?"

"It belongs to a man called Chada."

"Chada? Never heard of him. But I mean, what part of London is it in?"

"Whatever do you mean? It is in Limehouse, I believe. I don't understand. You came here."

"I didn't," said young Kerry cheerfully; "I was fetched!"

"By your father?"

"Not on your life. By a couple of Chinks! I'll tell you something." He raised his twinkling blue eyes. "We are properly up against it. I suppose you couldn't climb down a rain-pipe?"

CHAPTER 7

RETRIBUTION

It was that dark, still, depressing hour of the night, when all life is at its lowest ebb. In the low, strangely perfumed room of books Zani Chada sat before his table, his yellow hands clutching the knobs on his chair arms, his long, inscrutable eyes staring unseeingly before him.

Came a disturbance and the sound of voices, and Lou Chada, his son, stood at the doorway. He still wore his evening clothes, but he no longer looked smart. His glossy black hair was disheveled, and his handsome, olive face bore a hunted look. Panic was betoken by twitching mouth and fear-bright eyes. He stopped, glaring at his father, and:

"Why are you not gone?" asked the latter sternly. "Do you wish to wreck me as well as yourself ?"

"The police have posted a man opposite Kwee's house. I cannot get out that way."

"There was no one there when the boy was brought in."

"No, but there is now. Father!" He took a step forward. "I'm trapped. They sha'n't take me. You won't let them take me?"

Zani Chada stirred not a muscle, but:

"To-night," he said, "your mad passion has brought ruin to both of us. For the sake of a golden doll who is not worth the price of the jewels she wears, you have placed yourself within reach of the hangman."

"I was mad, I was mad," groaned the other.

"But I, who was sane, am involved in the consequences," retorted his father.

"He will be silent at the price of the boy's life."

"He may be," returned Zani Chada. "I hate him, but he is a man. Had you escaped, he might have consented to be silent. Once you are arrested, nothing would silence him."

"If the case is tried it will ruin Pat's reputation."

"What a pity!" said Zani Chada.

In some distant part of the house a gong was struck three times.

"Go," commanded his father. "Remain at Kwee's house until I send for you. Let Ah Fang go to the room above and see that the woman is silent. An outcry would ruin our last chance."

Lou Chada raised his hands, brushing the hair back from his wet forehead, then, staring haggardly at his father, turned and ran from the room.

A minute later Kerry was ushered in by the Chinese servant. The savage face was set like a mask. Without removing his hat, he strode across to the table and bent down so that fierce, wide- open blue eyes stared closely into long, half-closed black ones.

"I've got one thing to say," explained Kerry huskily. "Whatever the hangman may do to your slimy son, and whatever happens to the little blonde fool he kidnapped, if you've laid a hand on my kid I'll kick you to death, if I follow you round the world to do it."

Zani Chada made no reply, but his knuckles gleamed, so tightly did he clutch the knobs on the chair arms. Kerry's savagery would have awed any man, even though he had supposed it to be the idle threat of a passionate man. But Zani Chada knew all men, and he knew this one. When Daniel Kerry declared that in given circumstances he would kick Zani Chada to death, he did not mean that he would shoot him, strangle him, or even beat him with his fists; he meant precisely what he said--that he would kick him to death--and Zani Chada knew it.

Thus there were some moments of tense silence during which the savage face of the Chief Inspector drew even closer to the gaunt, yellow face of the Eurasian. Finally:

"Listen only for one moment," said Zani Chada. His voice had lost its guttural intonation. He spoke softly, sibilantly. "I, too, am a father..."

"Don't mince words!" shouted Kerry. "You've kidnapped my boy. If I have to tear your house down brick by brick I'll find him. And if you've hurt one hair of his head--you know what to expect!"

He quivered. The effort of suppression which he had imposed upon himself was frightful to witness. Zani Chada, student of men, knew that in despite of his own physical strength and of the hidden resources at his beck, he stood nearer to primitive retribution than he had ever done. Yet:

"I understand," he continued. "But you do not understand. Your boy is not in this house. Oh! violence cannot avail! It can only make his loss irreparable."

Kerry, nostrils distended, eyes glaring madly, bent over him.

"Your scallywag of a son," he said hoarsely, "has gone one step too far. His adventures have twice before ended in murder--and you have covered him. This time you can't do it. I'm not to be bought. We've stood for the Far East in London long enough. Your cub hangs this time. Get me? There'll be no bargaining. The woman's reputation won't stop me. My kid's danger won't stop me. But if you try to use him as a lever I'll boot you to your stinking yellow paradise and they'll check you in as pulp."

"You speak of three deaths," murmured Zani Chada.

Kerry clenched his teeth so tightly that his maxillary muscles protruded to an abnormal degree. He thrust his clenched fists into his coat pockets.

"We all follow our vocations in life," resumed the Eurasian, "to the best of our abilities. But is professional kudos not too dearly bought at the price of a loved one lost for ever? A far better bargain would be, shall we say, ten thousand pounds, as the price of a silk handkerchief..."

Kerry's fierce blue eyes closed for a fraction of a second. Yet, in that fraction of a second, he had visualized some of the things which ten thousand pounds--a sum he could never hope to possess--would buy. He had seen his home, as he would have it-- and he had seen Dan there, safe and happy at his mother's side. Was he entitled to disregard the happiness of his wife, the life of his boy, the honorable name of Sir Noel Rourke, because an outcast like Peters had come to a fitting end--because a treacherous Malay and a renegade Chinaman had, earlier, gone the same way, sped, as he suspected, by the same hand?

"My resources are unusual," added Chada, speaking almost in a whisper. "I have cash to this amount in my safe..."

So far he had proceeded when he was interrupted; and the cause of the interruption was this:

A few moments earlier another dramatic encounter had taken place in a distant part of the house. Kerry Junior, having scientifically tested all the possible modes of egress from the room in which Lady Pat was confined, had long ago desisted, and had exhausted his ingenuity in plans which discussion had proved to be useless. In spite of the novelty and the danger of his situation, nature was urging her laws. He was growing sleepy. The crowning tragedy had been the discovery that he could not regain the small, square window set high in the wall from which he had dropped into this luxurious prison. Now, as the two sat side by side upon a cushioned divan, the woman's arm about the boy's shoulders, they were startled to hear, in the depths of the house, three notes of a gong.

Young Kerry's sleepiness departed. He leapt to his feet as though electrified.

"What was that?"

There was something horrifying in those gong notes in the stillness of the night. Lady Pat's beautiful eyes grew glassy with fear.

"I don't know," replied Dan. "It seemed to come from below."

He ran to the door, drew the curtain aside, and pressed his ear against one of the panels, listening intently. As he did so, his attitude grew tense, his expression changed, then:

"We're saved!" he cried, turning a radiant face to the woman. "I heard my father's voice!"

"Oh, are you sure, are you sure?"

"Absolutely sure!"

He bent to press his ear to the panel again, when a stifled cry from his companion brought him swiftly to his feet. The second door in the room had opened silently, and a small Chinaman, who carried himself with a stoop, had entered, and now, a menacing expression upon his face, was quickly approaching the boy.

What he had meant to do for ever remained in doubt, for young Kerry, knowing his father to be in the house and seeing an open door before him, took matters into his own hands. At the moment that the silent Chinaman was about to throw his arms about him, the pride of the junior school registered a most surprising left accurately on the point of Ah Fang's jaw, following it up by a willful transgression of Queensberry rules in the form of a stomach

59

punch which temporarily decided the issue. Then:

"Quick! quick!" he cried breathlessly, grasping Lady Pat's hand. "This is where we run!"

In such fashion was Zani Chada interrupted, the interruption taking the form of a sudden, shrill outcry:

"Dad! dad! Where are you, dad?"

Kerry spun about as a man galvanized. His face became transfigured.

"This way, Dan!" he cried. "This way, boy!"

Came a clatter of hurrying feet, and into the low, perfumed room burst Dan Kerry, junior, tightly clasping the hand of a pale- faced, disheveled woman in evening dress. It was Lady Rourke; and although she seemed to be in a nearly fainting condition, Dan dragged her, half running, into the room.

Kerry gave one glance at the pair, then, instantly, he turned to face Zani Chada. The latter, like a man of stone, sat in his carved chair, eyes nearly closed. The Chief Inspector whipped out a whistle and raised it to his lips. He blew three blasts upon it.

From one--two--three--four points around the house the signal was answered.

Zani Chada fully opened his long, basilisk eyes.

"You win, Chief Inspector," he said. "But much may be done by clever counsel. If all fails..."

"Well?" rapped Kerry fiercely, at the same time throwing his arm around the boy.

"I may continue to take an interest in your affairs."

A tremendous uproar arose, within and without the house. The police were raiding the place. Lady Rourke sank down, slowly, almost at the Eurasian's feet.

But Chief Inspector Kerry experienced an unfamiliar chill as his uncompromising stare met the cold hatred which blazed out of the black eyes, narrowed, now, and serpentine, of Zani Chada.

THE PIGTAIL OF HI WING HO

CHAPTER I

HOW I OBTAINED IT

Leaving the dock gates behind me I tramped through the steady drizzle, going parallel with the river and making for the Chinese quarter. The hour was about half-past eleven on one of those September nights when, in such a locality as this, a stifling quality seems to enter the atmosphere, rendering it all but unbreathable. A mist floated over the river, and it was difficult to say if the rain was still falling, indeed, or if the ample moisture upon my garments was traceable only to the fog. Sounds were muffled, lights dimmed, and the frequent hooting of sirens from the river added another touch of weirdness to the scene.

Even when the peculiar duties of my friend, Paul Harley, called him away from England, the lure of this miniature Orient which I had first explored under his guidance, often called me from my chambers. In the house with the two doors in Wade Street, Limehouse, I would discard the armor of respectability, and, dressed in a manner unlikely to provoke comment in dockland, would haunt those dreary ways sometimes from midnight until close upon dawn. Yet, well as I knew the district and the strange and often dangerous creatures lurking in its many burrows, I experienced a chill partly physical and partly of apprehension to-night; indeed, strange though it may sound, I hastened my footsteps in order the sooner to reach the low den for which I was bound--Malay Jack's--a spot marked plainly on the crimes-map and which few respectable travelers would have regarded as a haven of refuge.

But the chill of the adjacent river, and some quality of utter desolation which seemed to emanate from the deserted wharves and ramshackle buildings about me, were driving me thither now; for I knew that human companionship, of a sort, and a glass of good liquor-- from a store which the Customs would have been happy to locate--awaited me there. I might chance, too, upon Durham or Wessex, of New Scotland Yard, both good friends of mine, or even upon the Terror of Chinatown, Chief Inspector Kerry, a man for whom I had an esteem which none of his ungracious manners could diminish.

I was just about to turn to the right into a narrow and nameless alley, lying at right angles to the Thames, when I pulled up sharply, clenching my fists and listening.

A confused and continuous sound, not unlike that which might be occasioned by several large and savage hounds at close grips, was proceeding out of the darkness ahead of me; a worrying, growling, and scuffling which presently I identified as human, although in fact it was animal enough. A moment I hesitated, then, distinguishing among the sounds of conflict an unmistakable, though subdued, cry for help, I leaped forward and found myself in the midst of the melee. This was taking place in the lee of a high, dilapidated brick wall. A lamp in a sort of iron bracket spluttered dimly above on the right, but the scene of the conflict lay in densest shadow, so that the figures were indistinguishable.

"Help! By Gawd! they're strangling me..."

From almost at my feet the cry arose and was drowned in Chinese chattering. But guided by it I now managed to make out that the struggle in progress waged between a burly English sailorman and two lithe Chinese. The yellow men seemed to have gained the advantage and my course was clear.

A straight right on the jaw of the Chinaman who was engaged in endeavoring to throttle the victim laid him prone in the dirty roadway. His companion, who was holding the wrist of the recumbent man, sprang upright as though propelled by a spring. I struck out at him savagely. He uttered a shrill scream not unlike that of a stricken hare, and fled so rapidly that he seemed to melt in the mist.

"Gawd bless you, mate!" came chokingly from the ground--and the rescued man, extricating himself from beneath the body of his stunned assailant, rose unsteadily to his feet and lurched toward me.

As I had surmised, he was a sailor, wearing a rough, blue-serge jacket and having his greasy trousers thrust into heavy seaboots--by which I judged that he was but newly come ashore. He stooped and picked up his cap. It was covered in mud, as were the rest of his garments, but he brushed it with his sleeve as though it had been but slightly soiled and clapped it on his head.

He grasped my hand in a grip of iron, peering into my face, and his breath was eloquent.

"I'd had one or two, mate," he confided huskily (the confession was unnecessary). "It was them two in the Blue Anchor as did it; if I 'adn't 'ad them last two, I could 'ave broke up them Chinks with one 'and tied behind me."

"That's all right," I said hastily, "but what are we going to do about this Chink here?" I added, endeavoring at the same time to extricate my hand from the vise-like grip in which he persistently held it. "He hit the tiles pretty heavy when he went down."

As if to settle my doubts, the recumbent figure suddenly arose and without a word fled into the darkness and was gone like a phantom. My new friend made no attempt to follow, but:

"You can't kill a bloody Chink," he confided, still clutching my hand; "it ain't 'umanly possible. It's easier to kill a cat. Come along o' me and 'ave one; then I'll tell you somethink. I'll put you on somethink, I will."

With surprising steadiness of gait, considering the liquid cargo he had aboard, the man, releasing my hand and now seizing me firmly by the arm, confidently led me by divers narrow ways, which I knew, to a little beerhouse frequented by persons of his class.

My own attire was such as to excite no suspicion in these surroundings, and although I considered that my acquaintance had imbibed more than enough for one night, I let him have his own way in order that I might learn the story which he seemed disposed to confide in me. Settled in the corner of the beerhouse--which chanced to be nearly empty-- with portentous pewters before us, the conversation was opened by my new friend:

"I've been paid off from the Jupiter--Samuelson's Planet Line," he explained. "What I am is a fireman."

"She was from Singapore to London?" I asked.

"She was," he replied, "and it was at Suez it 'appened--at Suez."

I did not interrupt him.

"I was ashore at Suez--we all was, owin' to a 'itch with the canal company--a matter of money, I may say. They make yer pay before they'll take yer through. Do you know that?"

I nodded.

"Suez is a place," he continued, "where they don't sell whisky, only poison. Was you ever at Suez?"

Again I nodded, being most anxious to avoid diverting the current of my friend's thoughts.

"Well, then," he continued, "you know Greek Jimmy's--and that's where I'd been."

I did not know Greek Jimmy's, but I thought it unnecessary to mention the fact.

"It was just about this time on a steamin' 'ot night as I come out of Jimmy's and started for the ship. I was walkin' along the Waghorn Quay, same as I might be walkin' along to-night, all by myself--bit of a list to port but nothing much--full o' joy an' happiness, 'appy an' free--'appy an' free. Just like you might have noticed to-night, I noticed a knot of Chinks scrappin' on the ground all amongst the dust right in front of me. I rammed in, windmillin' all round and knocking 'em down like skittles. Seemed to me there was about ten of 'em, but allowin' for Jimmy's whisky, maybe there wasn't more than three. Anyway, they all shifted and left me standin' there in the empty street with this 'ere in my 'and."

At that, without more ado, he thrust his hand deep into some concealed pocket and jerked out a Chinese pigtail, which had been severed, apparently some three inches from the scalp, by a clean cut. My acquaintance, with somewhat bleared eyes glistening in appreciation of his own dramatic skill--for I could not conceal my surprise--dangled it before me triumphantly.

"Which of 'em it belong to," he continued, thrusting it into another pocket and drumming loudly on the counter for more beer, "I can't say, 'cos I don't know. But that ain't all."

The tankards being refilled and my friend having sampled the contents of his own:

"That ain't all," he continued. "I thought I'd keep it as a sort of relic, like. What 'appened? I'll tell you. Amongst the crew there's three Chinks--see? We ain't through the canal before one of 'em, a new one to me--Li Ping is his name--offers me five bob for the pigtail, which he sees me looking at one mornin'. I give him a punch on the nose an' 'e don't renew the offer: but that night (we're layin' at Port Said) 'e tries to pinch it! I dam' near broke his neck, and 'e don't try any more. To-night"--he extended his right arm forensically--"a deppitation of Chinks waits on me at the dock gates; they explains as from a patriotic point of view they feels it to be their dooty to buy that pigtail off of me, and they bids a quid, a bar of gold--a Jimmy o' Goblin!"

He snapped his fingers contemptuously and emptied his pewter. A sense of what was coming began to dawn on me. That the "hold-up" near the riverside formed part of the scheme was possible, and, reflecting on my rough treatment of the two Chinamen, I chuckled inwardly. Possibly, however, the scheme had germinated in my acquaintance's mind merely as a result of an otherwise common assault, of a kind not unusual in these parts, but, whether elaborate or comparatively simple, that the story of the pigtail was a "plant" designed to reach my pocket, seemed a reasonable hypothesis.

"I told him to go to China," concluded the object of my suspicion, again rapping upon the counter, "and you see what come of it. All I got to say is this: If they're so bloody patriotic, I says one thing: I ain't the man to stand in their way. You done me a good turn to-night, mate; I'm doing you one. 'Ere's the bloody pigtail, 'ere's my empty mug. Fill the mug and the pigtail's yours. It's good for a quid at the dock gates any day!"

My suspicions vanished; my interest arose to boiling point. I refilled my acquaintance's mug, pressed a sovereign upon him (in honesty I must confess that he was loath to take it), and departed with the pigtail coiled neatly in an inner pocket of my jacket. I entered the house in Wade Street by the side door, and half an hour later let myself out by the front door, having cast off my dockland disguise.

CHAPTER 2

HOW I LOST IT

It was not until the following evening that I found leisure to examine my strange acquisition, for affairs of more immediate importance engrossed my attention. But at about ten o'clock I seated myself at my table, lighted the lamp, and taking out the pigtail from the table drawer, placed it on the blotting-pad and began to examine it with the greatest curiosity, for few Chinese affect the pigtail nowadays.

I had scarcely commenced my examination, however, when it was dramatically interrupted. The door bell commenced to ring jerkily. I stood up, and as I did so the ringing ceased and in its place came a muffled beating on the door. I hurried into the passage as the bell commenced ringing again, and I had almost reached the door when once more the ringing ceased; but now I could hear a woman's voice, low but agitated:

"Open the door! Oh, for God's sake be quick!"

Completely mystified, and not a little alarmed, I threw open the door, and in there staggered a woman heavily veiled, so that I could see little of her features, but by the lines of her figure I judged her to be young.

Uttering a sort of moan of terror she herself closed the door, and stood with her back to it, watching me through the thick veil, while her breast rose and fell tumultuously.

"Thank God there was someone at home!" she gasped.

I think I may say with justice that I had never been so surprised in my life; every particular of the incident marked it as unique--set it apart from the episodes of everyday life.

"Madam," I began doubtfully, "you seem to be much alarmed at something, and if I can be of any assistance to you..."

"You have saved my life!" she whispered, and pressed one hand to her bosom. "In a moment I will explain."

"Won't you rest a little after your evidently alarming experience?" I suggested.

My strange visitor nodded, without speaking, and I conducted her to the study which I had just left, and placed the most comfortable arm-chair close beside the table so that as I sat I might study this woman who so strangely had burst in upon me. I even tilted the shaded lamp, artlessly, a trick I had learned from Harley, in order that the light might fall upon her face.

She may have detected this device; I know not; but as if in answer to its challenge, she raised her gloved hands and unfastened the heavy veil which had concealed her features.

Thereupon I found myself looking into a pair of lustrous black eyes whose almond shape was that of the Orient; I found myself looking at a woman who, since she was evidently a Jewess, was probably no older than eighteen or nineteen, but whose beauty was ripely voluptuous, who might fittingly have posed for Salome, who, despite her modern fashionable garments, at once suggested to my mind the wanton beauty of the daughter of Herodias.

I stared at her silently for a time, and presently her full lips parted in a slow smile. My ideas were diverted into another channel.

"You have yet to tell me what alarmed you," I said in a low voice, but as courteously as possible, "and if I can be of any assistance in the matter."

My visitor seemed to recollect her fright--or the necessity for simulation. The pupils of her fine eyes seemed to grow larger and darker; she pressed her white teeth into her lower lips, and resting her hands upon the table leaned toward me.

"I am a stranger to London," she began, now exhibiting a certain diffidence, "and to-night I was looking for the chambers of Mr. Raphael Philips of Figtree Court."

"This is Figtree Court," I said, "but I know of no Mr. Raphael Philips who has chambers here."

The black eyes met mine despairingly.

"But I am positive of the address!" protested my beautiful but strange caller--from her left glove she drew out a scrap of paper, "here it is."

I glanced at the fragment, upon which, in a woman's hand the words were penciled: "Mr. Raphael Philips, 36-b Figtree Court, London."

I stared at my visitor, deeply mystified.

"These chambers are 36-b!" I said. "But I am not Raphael Philips, nor have I ever heard of him. My name is Malcolm Knox. There is evidently some mistake, but"--returning the slip of paper--"pardon me if I remind you, I have yet to learn the cause of your alarm."

"I was followed across the court and up the stairs."

"Followed! By whom?"

"By a dreadful-looking man, chattering in some tongue I did not understand!"

My amazement was momentarily growing greater.

"What kind of a man?" I demanded rather abruptly.

"A yellow-faced man--remember I could only just distinguish him in the darkness on the stairway, and see little more of him than his eyes at that, and his ugly gleaming teeth--oh! it was horrible!"

"You astound me," I said; "the thing is utterly incomprehensible." I switched off the light of the lamp. "I'll see if there's any sign of him in the court below."

"Oh, don't leave me! For heaven's sake don't leave me alone!"

She clutched my arm in the darkness.

"Have no fear; I merely propose to look out from this window."

Suiting the action to the word, I peered down into the court below. It was quite deserted. The night was a very dark one, and there were many patches of shadow in which a man might have lain concealed.

"I can see no one," I said, speaking as confidently as possible, and relighting the lamp, "if I call a cab for you and see you safely into it, you will have nothing to fear, I think."

"I have a cab waiting," she replied, and lowering the veil she stood up to go.

"Kindly allow me to see you to it. I am sorry you have been subjected to this annoyance, especially as you have not attained the object of your visit."

"Thank you so much for your kindness; there must be some mistake about the address, of course."

She clung to my arm very tightly as we descended the stairs, and often glanced back over her shoulder affrightedly, as we crossed the court. There was not a sign of anyone about, however, and I could not make up my mind whether the story of the yellow man was a delusion or a fabrication. I inclined to the latter theory, but the object of such a deception was more difficult to determine.

Sure enough, a taxicab was waiting at the entrance to the court; and my visitor, having seated herself within, extended her hand to me, and even through the thick veil I could detect her brilliant smile.

"Thank you so much, Mr. Knox," she said, "and a thousand apologies. I am sincerely sorry to have given you all this trouble."

The cab drove off. For a moment I stood looking after it, in a state of dreamy incertitude, then turned and slowly retraced my steps. Reopening the door of my chambers with my key, I returned to my study and sat down at the table to endeavor to arrange the facts of what I recognized to be a really amazing episode. The adventure, trifling though it seemed, undoubtedly held some hidden significance that at present was not apparent to me. In accordance with the excellent custom of my friend, Paul Harley, I prepared to make notes of the occurrence while the facts were still fresh in my memory. At the moment that I was about to begin, I made an astounding discovery.

Although I had been absent only a few minutes, and had locked my door behind me, the pigtail was gone!

I sat quite still, listening intently. The woman's story of the yellow man on the stairs suddenly assumed a totally different aspect--a new and sinister aspect. Could it be that the pigtail was at the bottom of the mystery?--could it be that some murderous Chinaman who had been lurking in hiding, waiting his opportunity, had in some way gained access to my chambers during that brief absence? If so, was he gone?

From the table drawer I took out a revolver, ascertained that it was fully loaded, and turning up light after light as I proceeded, conducted a room-to-room search. It was without result; there was absolutely nothing to indicate that anyone had surreptitiously entered or departed from my chambers.

I returned to the study and sat gazing at the revolver lying on the blotting-pad before me. Perhaps my mind worked slowly, but I think that fully fifteen minutes must have passed before it dawned on me that the explanation not only of the missing pigtail but of the other incidents of the night, was simple enough. The yellow man had been a fabrication, and my dark-eyed visitor had not been in quest of "Raphael Philips," but in quest of the pigtail: and her quest had been successful!

"What a hopeless fool I am!" I cried, and banged my fist down upon the table, "there was no yellow man at all--there was~~~~~"

My door bell rang. I sprang nervously to my feet, glanced at the revolver on the table--and finally dropped it into my coat pocket ere going out and opening the door.

On the landing stood a police constable and an officer in plain clothes.

"Your name is Malcolm Knox?" asked the constable, glancing at a note-book which he held in his hand.

"It is," I replied.

"You are required to come at once to Bow Street to identify a woman who was found murdered in a taxi-cab in the Strand about eleven o'clock to-night."

I suppressed an exclamation of horror; I felt myself turning pale.

"But what has it to do..."

"The driver stated she came from your chambers, for you saw her off, and her last words to you were 'Good night, Mr. Knox, I am sincerely sorry to have given you all this trouble.' Is that correct, sir?"

The constable, who had read out the information in an official voice, now looked at me, as I stood there stupefied.

"It is," I said blankly. "I'll come at once." It would seem that I had misjudged my unfortunate visitor: her story of the yellow man on the stair had apparently been not a fabrication, but a gruesome fact!

CHAPTER 3

HOW I REGAINED IT

My ghastly duty was performed; I had identified the dreadful thing, which less than an hour before had been a strikingly beautiful woman, as my mysterious visitor. The police were palpably disappointed at the sparsity of my knowledge respecting her. In fact, had it not chanced that Detective Sergeant Durham was in the station, I think they would have doubted the accuracy of my story.

As a man of some experience in such matters, I fully recognized its improbability, but beyond relating the circumstances leading up to my possession of the pigtail and the events which had ensued, I could do no more in the matter. The weird relic had not been found on the dead woman, nor in the cab.

Now the unsavory business was finished, and I walked along Bow Street, racking my mind for the master-key to this mystery in which I was become enmeshed. How I longed to rush off to Harley's rooms in Chancery Lane and to tell him the whole story! But my friend was a thousand miles away--and I had to see the thing out alone.

That the pigtail was some sacred relic stolen from a Chinese temple and sought for by its fanatical custodians was a theory which persistently intruded itself. But I could find no place in that hypothesis for the beautiful Jewess; and that she was intimately concerned I did not doubt. A cool survey of the facts rendered it fairly evident that it was she and none other who had stolen the pigtail from my rooms. Some third party--possibly the "yellow man" of whom she had spoken--had in turn stolen it from her, strangling her in the process.

The police theory of the murder (and I was prepared to accept it) was that the assassin had been crouching in hiding behind or beside the cab--or even within the dark interior. He had leaped in and attacked the woman at the moment that the taxi-man had started his engine; if already inside, the deed had proven even easier. Then, during some block in the traffic, he had slipped out unseen, leaving the body of the victim to be discovered when the cab pulled up at the hotel.

I knew of only one place in London where I might hope to obtain useful information, and for that place I was making now. It was Malay Jack's, whence I had been bound on the previous night when my strange meeting with the seaman who then possessed the pigtail had led to a change of plan. The scum of the Asiatic population always come at one time or another to Jack's, and I hoped by dint of a little patience to achieve what the police had now apparently despaired of achieving--the discovery of the assassin.

Having called at my chambers to obtain my revolver, I mounted an eastward-bound motor-bus. The night, as I have already stated, was exceptionally dark. There was no moon, and heavy clouds were spread over the sky; so that the deserted East End streets presented a sufficiently uninviting aspect, but one with which I was by no means unfamiliar and which certainly in no way daunted me.

Changing at Paul Harley's Chinatown base in Wade Street, I turned my steps in the same direction as upon the preceding night; but if my own will played no part in the matter, then decidedly Providence truly guided me. Poetic justice is rare enough in real life, yet I was destined to-night to witness swift retribution overtaking a malefactor.

The by-ways which I had trodden were utterly deserted; I was far from the lighted high road, and the only signs of human activity that reached me came from the adjacent river; therefore, when presently an outcry arose from somewhere on my left, for a moment I really believed that my imagination was vividly reproducing the episode of the night before!

A furious scuffle--between a European and an Asiatic--was in progress not twenty yards away!

Realizing that such was indeed the case, and that I was not the victim of hallucination, I advanced slowly in the direction of the sounds, but my footsteps reechoed hollowly from wall to wall of the narrow passage-way, and my coming brought the conflict to a sudden and dramatic termination.

"Thought I wouldn't know yer ugly face, did yer?" yelled a familiar voice. "No good squealin'--I got yer! I'd bust you up if I could!" (a sound of furious blows and inarticulate chattering) "but it ain't 'umanly possible to kill a Chink..."

I hurried forward toward the spot where two dim figures were locked in deadly conflict.

"Take that to remember me by!" gasped the husky voice as I ran up.

One of the figures collapsed in a heap upon the ground. The other made off at a lumbering gait along a second and even narrower passage branching at right angles from that in which the scuffle had taken place.

The clatter of the heavy sea-boots died away in the distance. I stood beside the fallen man, looking keenly about to right and left; for an impression was strong upon me that another than I had been witness of the scene--that a shadowy form had slunk back furtively at my approach. But the night gave up no sound in confirmation of this, and I could detect no sign of any lurker.

I stooped over the Chinaman (for a Chinaman it was) who lay at my feet, and directed the ray of my pocket-lamp upon his yellow and contorted countenance. I suppressed a cry of surprise and horror.

Despite the human impossibility referred to by the missing fireman, this particular Chinaman had joined the shades of his ancestors. I think that final blow, which had felled him, had brought his shaven skull in such violent contact with the wall that he had died of the thundering concussion set up.

Kneeling there and looking into his upturned eyes, I became aware that my position was not an enviable one, particularly since I felt little disposed to set the law on the track of the real culprit. For this man who now lay dead at my feet was doubtless one of the pair who had attempted the life of the fireman of the Jupiter.

That my seafaring acquaintance had designed to kill the Chinaman I did not believe, despite his stormy words: the death had been an accident, and (perhaps my morality was over-broad) I considered the assault to have been justified.

69

Now my ideas led me further yet. The dead Chinaman wore a rough blue coat, and gingerly, for I found the contact repulsive, I inserted my hand into the inside pocket. Immediately my fingers closed upon a familiar object--and I stood up, whistling slightly, and dangling in my left hand the missing pigtail!

Beyond doubt Justice had guided the seaman's blows. This was the man who had murdered my dark-eyed visitor!

I stood perfectly still, directing the little white ray of my flashlight upon the pigtail in my hand. I realized that my position, difficult before, now was become impossible; the possession of the pigtail compromised me hopelessly. What should I do?

"My God!" I said aloud, "what does it all mean?"

"It means," said a gruff voice, "that it was lucky I was following you and saw what happened!"

I whirled about, my heart leaping wildly. Detective-Sergeant Durham was standing watching me, a grim smile upon his face!

I laughed rather shakily.

"Lucky indeed!" I said. "Thank God you're here. This pigtail is a nightmare which threatens to drive me mad!"

The detective advanced and knelt beside the crumpled-up figure on the ground. He examined it briefly, and then stood up.

"The fact that he had the missing pigtail in his pocket," he said, "is proof enough to my mind that he did the murder."

"And to mine."

"There's another point," he added, "which throws a lot of light on the matter. You and Mr. Harley were out of town at the time of the Huang Chow case; but the Chief and I outlined it, you remember, one night in Mr. Harley's rooms?"

"I remember it perfectly; the giant spider in the coffin..."

"Yes; and a certain Ah Fu, confidential servant of the old man, who used to buy the birds the thing fed on. Well, Mr. Knox, Huang Chow was the biggest dealer in illicit stuff in all the East End--and this battered thing at our feet is--Ah Fu!"

"Huang Chow's servant?"

"Exactly!"

I stared, uncomprehendingly, and:

"In what way does this throw light on the matter?" I asked.

Durham--a very intelligent young officer--smiled significantly.

"I begin to see light!" he declared. "The gentleman who made off just as I arrived on the scene probably had a private quarrel with the Chinaman and was otherwise not concerned in any way."

"I am disposed to agree with you," I said guardedly.

"Of course, you've no idea of his identity?"

"I'm afraid not."

"We may find him," mused the officer, glancing at me shrewdly, "by applying at the offices of the Planet Line, but I rather doubt it. Also I rather doubt if we'll look very far. He's saved us a lot of trouble, but"--peering about in the shadowy corners which abounded--"didn't I see somebody else lurking around here?"

"I'm almost certain there was someone else!" I cried. "In fact, I could all but swear to it."

"H'm!" said the detective. "He's not here now. Might I trouble you to walk along to Limehouse Police Station for the ambulance? I'd better stay here."

I agreed at once, and started off.

Thus a second time my plans were interrupted, for my expedition that night ultimately led me to Bow Street, whence, after certain formalities had been observed, I departed for my chambers, the mysterious pigtail in my pocket. Failing the presence of Durham, the pigtail must have been retained as evidence, but:

"We shall know where to find it if it's wanted, Mr. Knox," said the Yard man, "and I can trust you to look after your own property."

The clock of St. Paul's was chiming the hour of two when I locked the door of my chambers and prepared to turn in. The clangor of the final strokes yet vibrated through the night's silence when someone set my own door bell loudly ringing.

With an exclamation of annoyance I shot back the bolts and threw open the door.

A Chinaman stood outside upon the mat!

CHAPTER 4

HOW IT ALL ENDED

"Me wishee see you," said the apparition, smiling blandly; "me comee in?"

"Come in, by all means," I said without enthusiasm, and, switching on the light in my study, I admitted the Chinaman and stood facing him with an expression upon my face which I doubt not was the reverse of agreeable.

My visitor, who wore a slop-shop suit, also wore a wide-brimmed bowler hat; now, the set bland smile still upon his yellow face, he removed the bowler and pointed significantly to his skull.

His pigtail had been severed some three inches from the root!

"You gotchee my pigtail," he explained; "me callee get it--thank you."

71

"Thank you," I said grimly. "But I must ask you to establish your claim rather more firmly."

"Yessir," agreed the Chinaman.

And thereupon in tolerable pidgin English he unfolded his tale. He proclaimed his name to be Hi Wing Ho, and his profession that of a sailor, or so I understood him. While ashore at Suez he had become embroiled with some drunken seamen: knives had been drawn, and in the scuffle by some strange accident his pigtail had been severed. He had escaped from the conflict, badly frightened, and had run a great distance before he realized his loss. Since Southern Chinamen of his particular Tong hold their pigtails in the highest regard, he had instituted inquiries as soon as possible, and had presently learned from a Chinese member of the crew of the S. S. Jupiter that the precious queue had fallen into the hands of a fireman on that vessel. He (Hi Wing Ho) had shipped on the first available steamer bound for England, having in the meanwhile communicated with his friend on the Jupiter respecting the recovery of the pigtail.

"What was the name of your friend on the Jupiter?"

"Him Li Ping--yessir!"--without the least hesitation or hurry.

I nodded. "Go on," I said.

He arrived at the London docks very shortly after the Jupiter. Indeed, the crew of the latter vessel had not yet been paid off when Hi Wing Ho presented himself at the dock gates. He admitted that, finding the fireman so obdurate, he and his friend Li Ping had resorted to violence, but he did not seem to recognize me as the person who had frustrated their designs. Thus far I found his story credible enough, excepting the accidental severing of the pigtail at Suez, but now it became wildly improbable, for he would have me believe that Li Ping, or Ah Fu, obtaining possession of the pigtail (in what manner Hi Wing Ho protested that he knew not) he sought to hold it to ransom, knowing how highly Hi Wing Ho valued it.

I glared sternly at the Chinaman, but his impassive countenance served him well. That he was lying to me I no longer doubted; for Ah Fu could not have hoped to secure such a price as would justify his committing murder; furthermore, the presence of the unfortunate Jewess in the case was not accounted for by the ingenious narrative of Hi Wing Ho. I was standing staring at him and wondering what course to adopt, when yet again my restless door-bell clamored in the silence.

Hi Wing Ho started nervously, exhibiting the first symptoms of alarm which I had perceived in him. My mind was made up in an instant. I took my revolver from the drawer and covered him.

"Be good enough to open the door, Hi Wing Ho," I said coldly.

He shrank from me, pouring forth voluble protestations.

"Open the door!"

I clenched my left fist and advanced upon him. He scuttled away with his odd Chinese gait and threw open the door. Standing before me I saw my friend Detective Sergeant Durham, and with him a remarkably tall and very large-boned man whose square-jawed face was deeply tanned and whose aspect was dourly Scottish.

When the piercing eyes of this stranger rested upon Hi Wing Ho an expression which I shall never forget entered into them; an expression coldly murderous. As for the Chinaman, he literally crumpled up.

"You rat!" roared the stranger.

Taking one long stride he stooped upon the Chinaman, seized him by the back of the neck as a terrier might seize a rat, and lifted him to his feet.

"The mystery of the pigtail, Mr. Knox," said the detective, "is solved at last."

"Have ye got it?" demanded the Scotsman, turning to me, but without releasing his hold upon the neck of Hi Wing Ho.

I took the pigtail from my pocket and dangled it before his eyes.

"Suppose you come into my study," I said, "and explain matters."

We entered the room which had been the scene of so many singular happenings. The detective and I seated ourselves, but the Scotsman, holding the Chinaman by the neck as though he had been some inanimate bundle, stood just within the doorway, one of the most gigantic specimens of manhood I had ever set eyes upon.

"You do the talking, sir," he directed the detective; "ye have all the facts."

While Durham talked, then, we all listened--excepting the Chinaman, who was past taking an intelligent interest in anything, and who, to judge from his starting eyes, was being slowly strangled.

"The gentleman," said Durham--"Mr. Nicholson--arrived two days ago from the East. He is a buyer for a big firm of diamond merchants, and some weeks ago a valuable diamond was stolen from him..."

"By this!" interrupted the Scotsman, shaking the wretched Hi Wing Ho terrier fashion.

"By Hi Wing Ho," explained the detective, "whom you see before you. The theft was a very ingenious one, and the man succeeded in getting away with his haul. He tried to dispose of the diamond to a certain Isaac Cohenberg, a Singapore moneylender; but Isaac Cohenberg was the bigger crook of the two. Hi Wing Ho only escaped from the establishment of Cohenberg by dint of sandbagging the moneylender, and quitted the town by a boat which left the same night. On the voyage he was indiscreet enough to take the diamond from its hiding-place and surreptitiously to examine it. Another member of the Chinese crew, one Li Ping-- otherwise Ah Fu, the accredited agent of old Huang Chow!--was secretly watching our friend, and, knowing that he possessed this valuable jewel, he also learned where he kept it hidden. At Suez Ah Fu attacked Hi Wing Ho and secured possession of the diamond. It was to secure possession of the diamond that Ah Fu had gone out East. I don't doubt it. He employed Hi Wing Ho--and Hi Wing Ho tried to double on him!

"We are indebted to you, Mr. Knox, for some of the data upon which we have reconstructed the foregoing and also for the next link in the narrative. A fireman ashore from the Jupiter intruded upon the scene at Suez and deprived Ah Fu of the fruits of his labors. Hi Wing Ho seems to have been badly damaged in the scuffle, but Ah Fu, the more wily of the two, evidently followed the fireman, and, deserting from his own ship, signed on with the Jupiter."

While this story was enlightening in some respects, it was mystifying in others. I did not interrupt, however, for Durham immediately resumed:

"The drama was complicated by the presence of a fourth character--the daughter of Cohenberg. Realizing that a small fortune had slipped through his fingers, the old moneylender dispatched his daughter in pursuit of Hi Wing Ho, having learned upon which vessel the latter had sailed. He had no difficulty in obtaining this information, for he is in touch with all the crooks of the town. Had he known that the diamond had been stolen by an agent of Huang Chow, he would no doubt have hesitated. Huang Chow has an international reputation.

"However, his daughter--a girl of great personal beauty--relied upon her diplomatic gifts to regain possession of the stone, but, poor creature, she had not counted with Ah Fu, who was evidently watching your chambers (while Hi Wing Ho, it seems, was assiduously shadowing Ah Fu!). How she traced the diamond from point to point of its travels we do not know, and probably never shall know, but she was undeniably clever and unscrupulous. Poor girl! She came to a dreadful end. Mr. Nicholson, here, identified her at Bow Street to-night."

Now the whole amazing truth burst upon me.

"I understand!" I cried. "This"--and I snatched up the pigtail--

"That my pigtail," moaned Hi Wing Ho feebly.

Mr. Nicholson pitched him unceremoniously into a corner of the room, and taking the pigtail in his huge hand, clumsily unfastened it. Out from the thick part, some two inches below the point at which it had been cut from the Chinaman's head, a great diamond dropped upon the floor!

For perhaps twenty seconds there was perfect silence in my study. No one stooped to pick the diamond from the floor--the diamond which now had blood upon it. No one, so far as my sense informed me, stirred. But when, following those moments of stupefaction, we all looked up--Hi Wing Ho, like a phantom, had faded from the room!

THE HOUSE OF GOLDEN JOSS

CHAPTER I

THE BLOOD-STAINED IDOL

"Stop when we pass the next lamp and give me a light for my pipe."

"Why?"

"No! don't look round," warned my companion. "I think someone is following us. And it is always advisable to be on guard in this neighborhood."

We had nearly reached the house in Wade Street, Limehouse, which my friend used as a base for East End operations. The night was dark but clear, and I thought that presently when dawn came it would bring a cold, bright morning. There was no moon, and as we passed the lamp and paused we stood in almost total darkness.

Facing in the direction of the Council School I struck a match. It revealed my ruffianly looking companion--in whom his nearest friends must have failed to recognize Mr. Paul Harley of Chancery Lane.

He was glancing furtively back along the street, and when a moment later we moved on, I too, had detected the presence of a figure stumbling toward us.

"Don't stop at the door," whispered Harley, for our follower was only a few yards away.

Accordingly we passed the house in which Harley had rooms, and had proceeded some fifteen paces farther when the man who was following us stumbled in between Harley and myself, clutching an arm of either. I scarcely knew what to expect, but was prepared for anything, when:

"Mates!" said a man huskily. "Mates, if you know where I can get a drink, take me there!"

Harley laughed shortly. I cannot say if he remained suspicious of the newcomer, but for my own part I had determined after one glance at the man that he was merely a drunken fireman newly recovered from a prolonged debauch.

"Where 'ave yer been, old son?" growled Harley, in that wonderful dialect of his which I had so often and so vainly sought to cultivate. "You look as though you'd 'ad one too many already."

"I ain't," declared the fireman, who appeared to be in a semi- dazed condition. "I ain't 'ad one since ten o'clock last night. It's dope wot's got me, not rum."

"Dope!" said Harley sharply; "been 'avin' a pipe, eh?"

"If you've got a corpse-reviver anywhere," continued the man in that curious, husky voice, "'ave pity on me, mate. I seen a thing to-night wot give me the jim-jams."

"All right, old son," said my friend good-humouredly; "about turn! I've got a drop in the bottle, but me an' my mate sails to-morrow, an' it's the last."

"Gawd bless yer!" growled the fireman; and the three of us--an odd trio, truly--turned about, retracing our steps.

As we approached the street lamp and its light shone upon the haggard face of the man walking between us, Harley stopped, and:

"Wot's up with yer eye?" he inquired.

He suddenly tilted the man's head upward and peered closely into one of his eyes. I suppressed a gasp of surprise for I instantly recognized the fireman of the Jupiter!

"Nothin' up with it, is there?" said the fireman.

"Only a lump o' mud," growled Harley, and with a very dirty handkerchief he pretended to remove the imaginary stain, and then, turning to me:

"Open the door, Jim," he directed.

His examination of the man's eyes had evidently satisfied him that our acquaintance had really been smoking opium.

We paused immediately outside the house for which we had been bound, and as I had the key I opened the door and the three of us stepped into a little dark room. Harley closed the door and we stumbled upstairs to a low first-floor apartment facing the street. There was nothing in its appointments, as revealed in the light of an oil lamp burning on the solitary table, to distinguish it from a thousand other such apartments which may be leased for a few shillings a week in the neighborhood. That adjoining might have told a different story, for it more closely resembled an actor's dressing-room than a seaman's lodging; but the door of this sanctum was kept scrupulously locked.

"Sit down, old son," said my friend heartily, pushing forward an old arm-chair. "Fetch out the grog, Jim; there's about enough for three."

I walked to a cupboard, as the fireman sank limply down in the chair, and took out a bottle and three glasses. When the man, who, as I could now see quite plainly, was suffering from the after effects of opium, had eagerly gulped the stiff drink which I handed to him, he looked around with dim, glazed eyes, and:

"You've saved my life, mates," he declared. "I've 'ad a 'orrible nightmare, I 'ave--a nightmare. See?"

He fixed his eyes on me for a moment, then raised himself from his seat, peering narrowly at me across the table.

"I seed you before, mate. Gaw, blimey! if you ain't the bloke wot I giv'd the pigtail to! And wot laid out that blasted Chink as was scraggin' me! Shake, mate!"

I shook hands with him, Harley eyeing me closely the while, in a manner which told me that his quick brain had already supplied the link connecting our doped acquaintance with my strange experience during his absence. At the same time it occurred to me that my fireman friend did not know that Ah Fu was dead, or he would never have broached the subject so openly.

"That's so," I said, and wondered if he required further information.

"It's all right, mate. I don't want to 'ear no more about blinking pigtails--not all my life I don't," and he sat back heavily in his chair and stared at Harley.

"Where have you been?" inquired Harley, as if no interruption had occurred, and then began to reload his pipe: "at Malay Jack's or at Number Fourteen?"

"Neither of 'em!" cried the fireman, some evidence of animation appearing in his face; "I been at Kwen Lung's."

"In Pennyfields?"

"That's 'im, the old bloke with the big joss. I allers goes to see Ma Lorenzo when I'm in Port o' London. I've seen 'er for the last time, mates."

He banged a big and dirty hand upon the table.

"Last night I see murder done, an' only that I know they wouldn't believe me, I'd walk across to Limehouse P'lice Station presently and put the splits on 'em, I would."

Harley, who was seated behind the speaker, glanced at me significantly.

"Sure you wasn't dreamin'?" he inquired facetiously.

"Dreamin'!" cried the man. "Dreams don't leave no blood be'ind, do they?"

"Blood!" I exclaimed.

"That's wot I said--blood! When I woke up this mornin' there was blood all on that grinnin' joss--the blood wot 'ad dripped from 'er shoulders when she fell."

"Eh!" said Harley. "Blood on whose shoulders? Wot the 'ell are you talkin' about, old son?"

"'Ere"--the fireman turned in his chair and grasped Harley by the arm--"listen to me, and I'll tell you somethink, I will. I'm goin' in the Seahawk in the mornin' see? But if you want to know somethink, I'll tell yer. Drunk or sober I bars the blasted p'lice, but if you like to tell 'em I'll put you on somethink worth tellin'. Sure the bottle's empty, mates?"

I caught Harley's glance and divided the remainder of the whisky evenly between the three glasses.

"Good 'ealth," said the fireman, and disposed of his share at a draught. "That's bucked me up wonderful."

He lay back in his chair and from a little tobacco-box began to fill a short clay pipe.

"Look 'ere, mates, I'm soberin' up, like, after the smoke, an' I can see, I can see plain, as nobody'll ever believe me. Nobody ever does, worse luck, but 'ere goes. Pass the matches."

He lighted his pipe, and looking about him in a sort of vaguely aggressive way:

"Last night," he resumed, "after I was chucked out of the Dock Gates, I made up my mind to go and smoke a pipe with old Ma Lorenzo. Round I goes to Pennyfields, and she don't seem glad to see me. There's nobody there only me. Not like the old days when you 'ad to book your seat in advance."

77

He laughed gruffly.

"She didn't want to let me in at first, said they was watched, that if a Chink 'ad an old pipe wot 'ad b'longed to 'is grandfather it was good enough to get 'im fined fifty quid. Anyway, me bein' an old friend she spread a mat for me and filled me a pipe. I asked after old Kwen Lung, but, of course, 'e was out gamblin', as usual; so after old Ma Lorenzo 'ad made me comfortable an' gone out I 'ad the place to myself, and presently I dozed off and forgot all about bloody ship's bunkers an' nigger-drivin' Scotchmen."

He paused and looked about him defiantly.

"I dunno 'ow long I slept," he continued, "but some time in the night I kind of 'alf woke up."

At that he twisted violently in his chair and glared across at Harley:

"You been a pal to me," he said; "but tell me I was dreamin' again and I'll smash yer bloody face!"

He glared for a while, then addressing his narrative more particularly to me, he resumed:

"It was a scream wot woke me--a woman's scream. I didn't sit up; I couldn't. I never felt like it before. It was the same as bein' buried alive, I should think. I could see an' I could 'ear, but I couldn't move one muscle in my body. Foller me? An' wot did I see, mates, an' wot did I 'ear? I'm goin' to tell yer. I see old Kwen Lung's daughter..."

"I didn't know 'e 'ad one," murmured Harley.

"Then you don't know much!" shouted the fireman. "I knew years ago, but 'e kept 'er stowed away somewhere up above, an' last night was the first time I ever see 'er. It was 'er shriek wot 'ad reached me, reached me through the smoke. I don't take much stock in Chink gals in general, but this one's mother was no Chink, I'll swear. She was just as pretty as a bloomin' ivory doll, an' as little an' as white, and that old swine Kwen Lung 'ad tore the dress off of 'er shoulders with a bloody great whip!"

Harley was leaning forward in his seat now, intent upon the man's story, and although I could not get rid of the idea that our friend was relating the events of a particularly unpleasant opium dream, nevertheless I was fascinated by the strange story and by the strange manner of its telling.

"I saw the blood drip from 'er bare shoulders, mates," the man continued huskily, and with his big dirty hands he strove to illustrate his words. "An' that old yellow devil lashed an' lashed until the poor gal was past screamin'. She just sunk down on the floor all of a 'cap, moanin' and moanin'--Gawd! I can 'ear 'er moanin' now!"

"Meanwhile, 'ere's me with murder in me 'eart lyin' there watchin', an' I can't speak, no! I can't even curse the yellow rat, an' I can't move--not a 'and, not a foot! Just as she fell there right up against the joss an' 'er blood trickled down on 'is gilded feet, old Ma Lorenzo comes staggerin' in. I remember all this as clear as print, mates, remember it plain, but wot 'appened next ain't so good an' clear. Somethink seemed to bust in me 'ead. Only just before I went off, the winder--there's only one in the room--was smashed to smithereens an' somebody come in through it."

"Are you sure?" said Harley eagerly. "Are you sure?"

That he was intensely absorbed in the story he revealed by a piece of bad artistry, very rare in him. He temporarily forgot his dialect. Our marine friend, however, was too much taken up with his own story to notice the slip, and:

"Dead sure!" he shouted.

He suddenly twisted around in his chair.

"Tell me I was dreamin', mate," he invited, "and if you ain't dreamin' in 'arf a tick it won't be because I 'aven't put yer to sleep!"

"I ain't arguin', old son," said Harley soothingly. "Get on with your yarn."

"Ho!" said the fireman, mollified, "so long as you ain't. Well, then, it's all blotted out after that. Somebody come in at the winder, but 'oo it was or wot it was I can't tell yer, not for fifty quid. When I woke up, which is about 'arf an hour before you see me, I'm all alone--see? There's no sign of Kwen Lung nor the gal nor old Ma Lorenzo nor anybody. I sez to meself, wot you keep on sayin'. I sez, 'You're dreamin', Bill.'"

"But I don't think you was," declared Harley. "Straight I don't."

"I know I wasn't!" roared the fireman, and banged the table lustily. "I see 'er blood on the joss an' on the floor where she lay!"

"This morning?" I interjected.

"This mornin', in the light of the little oil lamp where old Ma Lorenzo 'ad roasted the pills! It's all still an' quiet an' I feel more dead than alive. I'm goin' to give 'er a hail, see? When I sez to myself, 'Bill,' I sez, 'put out to sea; you're amongst Kaffirs, Bill.' It occurred to me as old Kwen Lung might wonder 'ow much I knew. So I beat it. But when I got in the open air I felt I'd never make my lodgin's without a tonic. That's 'ow I come to meet you, mates.

"Listen--I'm away in the old Seahawk in the mornin', but I'll tell you somethink. That yellow bastard killed his daughter last night! Beat 'er to death. I see it plain. The sweetest, prettiest bit of ivory as Gawd ever put breath into. If 'er body ain't in the river, it's in the 'ouse. Drunk or sober, I never could stand the splits, but mates"--he stood up, and grasping me by the arm, he drew me across the room where he also seized Harley in his muscular grip--"mates," he went on earnestly, "she was the sweetest, prettiest little gal as a man ever clapped eyes on. One of yer walk into Limehouse Station an' put the koppers wise. I'd sleep easier at sea if I knew old Kwen Lung 'ad gone west on a bloody rope's end."

CHAPTER 2

AT KWEN LUNG'S

For fully ten minutes after the fireman had departed Paul Harley sat staring abstractedly in front of him, his cold pipe between his teeth, and knowing his moods I intruded no words upon this reverie, until:

"Come on, Knox," he said, standing up suddenly, "I think this matter calls for speedy action."

"What! Do you think the man's story was true?"

"I think nothing. I am going to look at Kwen Lung's joss."

Without another word he led the way downstairs and out into the deserted street. The first gray halftones of dawn were creeping into the sky, so that the outlines of Limehouse loomed like dim silhouettes about us. There was abundant evidence in the form of noises, strange and discordant, that many workers were busy on dock and riverside, but the streets through which our course lay were almost empty. Sometimes a furtive shadow would move out of some black gully and fade into a dimly seen doorway in a manner peculiarly unpleasant and Asiatic. But we met no palpable pedestrian throughout the journey.

Before the door of a house in Pennyfields which closely resembled that which we had left in Wade Street, in that it was flatly uninteresting, dirty and commonplace, we paused. There was no sign of life about the place and no lights showed at any of the windows, which appeared as dim cavities--eyeless sockets in the gray face of the building, as dawn proclaimed the birth of a new day.

Harley seized the knocker and knocked sharply. There was no response, and he repeated the summons, but again without effect. Thereupon, with a muttered exclamation, he grasped the knocker a third time and executed a veritable tattoo upon the door. When this had proceeded for about half a minute or more:

"All right, all right!" came a shaky voice from within. "I'm coming."

Harley released the knocker, and, turning to me:

"Ma Lorenzo," he whispered. "Don't make any mistakes."

Indeed, even as he warned me, heralded by a creaking of bolts and the rattling of a chain, the door was opened by a fat, shapeless, half-caste woman of indefinite age; in whose dark eyes, now sunken in bloated cheeks, in whose full though drooping lips, and even in the whole overlaid contour of whose face and figure it was possible to recognize the traces of former beauty. This was Ma Lorenzo, who for many years had lived at that address with old Kwen Lung, of whom strange stories were told in Chinatown.

As Bill Jones, A.B., my friend, Paul Harley, was well known to Ma Lorenzo as he was well known to many others in that strange colony which clusters round the London docks. I sometimes enjoyed the privilege of accompanying my friend on a tour of investigation through the weird resorts which abound in that neighborhood, and, indeed, we had been returning from one of these Baghdad nights when our present adventure had been thrust upon us. Assuming a wild and boisterous manner which he had at command:

"'Urry up, Ma!" said Harley, entering without ceremony; "I want to introduce my pal Jim 'ere to old Kwen Lung, and make it all right for him before I sail."

Ma Lorenzo, who was half Portuguese, replied in her peculiar accent:

"This no time to come waking me up out of bed!"

But Harley, brushing past her, was already inside the stuffy little room, and I hastened to follow.

"Kwen Lung!" shouted my friend loudly. "Where are you? Brought a friend to see you."

"Kwen Lung no hab," came the complaining tones of Ma Lorenzo from behind us.

It was curious to note how long association with the Chinese had resulted in her catching the infection of that pidgin-English which is a sort of esperanto in all Asiatic quarters.

"Eh!" cried my friend, pushing open a door on the right of the passage and stumbling down three worn steps into a very evil- smelling room. "Where is he?"

"Go play fan-tan. Not come back."

Ma Lorenzo, having relocked the street door, had rejoined us, and as I followed my friend down into the dim and uninviting apartment she stood at the top of the steps, hands on hips, regarding us.

The place, which was quite palpably an opium den, must have disappointed anyone familiar with the more ornate houses of Chinese vice in San Francisco and elsewhere. The bare floor was not particularly clean, and the few decorations which the room boasted were garishly European for the most part. A deep divan, evidently used sometimes as a bed, occupied one side of the room, and just to the left of the steps reposed the only typically Oriental object in the place. It was a strange thing to see in so sordid a setting; a great gilded joss, more than life-size, squatting, hideous, upon a massive pedestal; a figure fit for some native temple but strangely out of place in that dirty little Limehouse abode.

I had never before visited Kwen Lung's, but the fame of his golden joss had reached me, and I know that he had received many offers for it, all of which he had rejected. It was whispered that Kwen Lung was rich, that he was a great man among the Chinese, and even that some kind of religious ceremony periodically took place in his house. Now, as I stood staring at the famous idol, I saw something which made me stare harder than ever.

The place was lighted by a hanging lamp from which depended bits of colored paper and several gilded silk tassels; but dim as the light was it could not conceal those tell-tale stains.

There was blood on the feet of the golden idol!

All this I detected at a glance, but ere I had time to speak:

"You can't tell me that tale, Ma!" cried Harley. "I believe 'e was smokin' in 'ere when we knocked."

The woman shrugged her fat shoulders.

"No, hab," she repeated. "You two johnnies clear out. Let me sleep."

But as I turned to her, beneath the nonchalant manner I could detect a great uneasiness; and in her dark eyes there was fear. That Harley also had seen the bloodstains I was well aware, and I did not doubt that furthermore he had noted the fact that the only mat which the room boasted had been placed before the joss-- doubtless to hide other stains upon the boards.

As we stood so I presently became aware of a current of air passing across the room in the direction of the open door. It came from a window before which a tawdry red curtain had been draped. Either the window behind the curtain was wide open, which is alien to Chinese habits, or it was shattered. While I was wondering if Harley intended to investigate further:

"Come on, Jim!" he cried boisterously, and clapped me on the shoulder; "the old fox don't want to be disturbed."

He turned to the woman:

"Tell him when he wakes up, Ma," he said, "that if ever my pal Jim wants a pipe he's to 'ave one. Savvy? Jim's square."

"Savvy," replied the woman, and she was wholly unable to conceal her relief. "You clear out now, and I tell Kwen Lung when he come in."

"Righto, Ma!" said Harley. "Kiss 'im on both cheeks for me, an' tell 'im I'll be 'ome again in a month."

Grasping me by the arm he lurched up the steps, and the two of us presently found ourselves out in the street again. In the growing light the squalor of the district was more evident than ever, but the comparative freshness of the air was welcome after the reek of that room in which the golden idol sat leering, with blood at his feet.

"You saw, Harley?" I exclaimed excitedly. "You saw the stains? And I'm certain the window was broken!"

Harley nodded shortly.

"Back to Wade Street!" he said. "I allow myself fifteen minutes to shed Bill Jones, able seaman, and to become Paul Harley, of Chancery Lane."

As we hurried along:

"What steps shall you take?" I asked.

"First step: search Kwen Lung's house from cellar to roof. Second step: entirely dependent upon result of first. The Chinese are subtle, Knox. If Kwen Lung has killed his daughter, it may require all the resources of Scotland Yard to prove it."

"But..."

"There is no 'but' about it. Chinatown is the one district of London which possesses the property of swallowing people up."

CHAPTER 3

"CAPTAIN DAN"

Half an hour later, as I sat in the inner room before the great dressing-table laboriously removing my disguise--for I was utterly incapable of metamorphosing myself like Harley in seven minutes--I heard a rapping at the outer door. I glanced nervously at my face in the mirror.

Comparatively little of "Jim" had yet been removed, for since time was precious to my friend I had acted as his dresser before setting to work to remove my own make-up. There were two entrances to the establishment, by one of which Paul Harley invariably entered and invariably went out, and from the other of which "Bill Jones" was sometimes seen to emerge, but never Paul Harley. That my friend had made good his retirement I knew, but, nevertheless, if I had to open the door of the outer room it must be as "Jim."

Thinking it impolite not to do so, since the one who knocked might be aware that we had come in but not gone out again, I hastily readjusted that side of my moustache which I had begun to remove, replaced my cap and muffler, and carefully locking the door of the dressing-room, crossed the outer apartment and opened the door.

It was Harley's custom never to enter or leave these rooms except under the mantle of friendly night, but at so early an hour I confess I had not expected a visitor. Wondering whom I should find there I opened the door.

Standing on the landing was a fellow-lodger who permanently occupied the two top rooms of the house. Paul Harley had taken the trouble to investigate the man's past, for "Captain Dan," the name by which he was known in the saloons and worse resorts which he frequented, was palpably a broken-down gentleman; a piece of flotsam caught in the yellow stream. Opium had been his downfall. How he lived I never knew, but Harley believed he had some small but settled income, sufficient to enable him to kill himself in comfort with the black pills.

As he stood there before me in the early morning light, I was aware of some subtle change in his appearance. It was fully six months since I had seen him last, but in some vague way he looked younger. Haggard he was, with an ugly cut showing on his temple, but not so lined as I remembered him. Some former man seemed to be struggling through the opium-scarred surface. His eyes were brighter, and I noted with surprise that he wore decent clothes and was clean shaved.

"Good morning, Jim," he said; "you remember me, don't you?"

As he spoke I observed, too, that his manner had altered. He who had consorted with the sweepings af the doss-houses now addressed me as a courteous gentleman addresses an inferior--not haughtily or patronizingly, but with a note of conscious superiority and self-respect wholly unfamiliar. Almost it threw me off my guard, but remembering in the nick of time that I was still "Jim":

"Of course I remember you, Cap'n," I said. "Step inside."

"Thanks," he replied, and followed me into the little room.

I placed for him the arm-chair which our friend the fireman had so recently occupied, but:

"I won't sit down," he said.

And now I observed that he was evidently in a condition of repressed excitement. Perhaps he saw the curiosity in my glance, for he suddenly rested both his hands on my shoulders, and:

"Yes, I have given up the dope, Jim," he said---"done with it for ever. There's not a soul in this neighborhood I can trust, yet if ever a man wanted a pal, I want one to-day. Now, you're square, my lad. I always knew that, in spite of the dope; and if I ask you to do a little thing that means a lot to me, I think you will do it. Am I right?"

"If it can be done, I'll do it," said I.

"Then, listen. I'm leaving England in the Patna for Singapore. She sails at noon to-morrow, and passengers go on board at ten o'clock. I've got my ticket, papers in order, but"--he paused impressively, grasping my shoulders hard--"I must get on board to-night."

I stared him in the face.

"Why?" I asked.

He returned my look with one searching and eager; then:

"If I show you the reason," said he, "and trust you with all my papers, will you go down to the dock--it's no great distance-- and ask to see Marryat, the chief officer? Perhaps you've sailed with him?"

"No," I replied guardedly. "I was never in the Patna."

"Never mind. When you give him a letter which I shall write he will make the necessary arrangements for me to occupy my state- room to-night. I knew him well," he explained, "in--the old days. Will you do it, Jim?"

"I'll do it with pleasure," I answered.

"Shake!" said Captain Dan.

We shook hands heartily, and:

"Now I'll show you the reason," he added. "Come upstairs."

Turning, he led the way upstairs to his own room, and wondering greatly, I followed him in. Never having been in Captain Dan's apartments I cannot say whether they, like their occupant, had changed for the better. But I found myself in a room surprisingly clean and with a note of culture in its appointments which was even more surprising.

On a couch by the window, wrapped in a fur rug, lay the prettiest half-caste girl I had ever seen, East or West. Her skin was like cream rose petals and her abundant hair was of wonderful lustrous black. Perhaps it was her smooth warm color which suggested the idea, but as her cheeks flushed at sight of Captain Dan and the long dark eyes lighted up in welcome, I thought of a delicate painting on ivory and I wondered more and more what it all could mean.

"I have brought Jim to see you," said Captain Dan. "No, don't trouble to move dear."

But even before he had spoken I had seen the girl wince with pain as she had endeavored to sit up to greet us. She lay on her side in a rather constrained attitude, but although her sudden movement had brought tears to her eyes she smiled bravely and extended a tiny ivory hand to me.

"This is my wife, Jim!" said Captain Dan.

I could find no words at all, but merely stood there looking very awkward and feeling almost awed by the indescribable expression of trust in the eyes of the little Eurasian, as with her tiny fingers hidden in her husband's clasp she lay looking up at him.

"Now you know, Jim," said he, "why we must get aboard the Patna to-night. My wife is really too ill to travel; in fact, I shall have to carry her down to the cab, and such a proceeding in daylight would attract an enormous crowd in this neighborhood!"

"Give me the letters and the papers," I answered. "I will start now."

His wife disengaged her hand and extended it to me.

"Thank you," she said, in a queer little silver-bell voice; "you are good. I shall always love you."

CHAPTER 4

THE SECRET OF MA LORENZO

It must have been about eleven o'clock that night when Paul Harley rang me up. Since we had parted in the early morning I had had no word from him, and I was all anxiety to tell him of the quaint little romance which unknown to us had had its setting in the room above.

In accordance with my promise I had seen the chief officer of the Patna; and from the start of surprise which he gave on opening "Captain Dan's" letter, I judged that Mr. Marryat and the man who for so long had sunk to the lowest rung of the ladder had been close friends in those "old days." At any rate, he had proceeded to make the necessary arrangements without a moment's delay, and the couple were to go on board the Patna at nine o'clock.

It was with a sense of having done at least one good deed that I finally quitted our Limehouse base and returned to my rooms. Now, at eleven o'clock at night:

"Can you come round to Chancery Lane at once?" said Harley. "I want you to run down to Pennyfields with me."

"Some development in the Kwen Lung business?"

"Hardly a development, but I'm not satisfied, Knox. I hate to be beaten."

Twenty minutes later I was sitting in Harley's study, watching him restlessly promenading up and down before the fire.

"The police searched Kwen Lung's place from foundation to tiles," he said. "I was there myself. Old Kwen Lung conveniently kept out of the way--still playing fan-tan, no doubt! But Ma Lorenzo was in evidence. She blandly declared that Kwen Lung never had a daughter! And in the absence of our friend the fireman, who sailed in the Seahawk, and whose evidence, by the way, is legally valueless--what could we do? They could find nobody in the neighborhood prepared to state that Kwen Lung had a daughter or that Kwen Lung had no daughter. There are all sorts of fables about the old fox, but the facts about him are harder to get at."

"But," I explained, "the bloodstains on the joss!"

"Ma Lorenzo stumbled and fell there on the previous night, striking her skull against the foot of the figure."

"What nonsense!" I cried. "We should have seen the wound last night."

"We might have done," said Harley musingly; "I don't know when she inflicted it on herself; but I did see it this morning."

"What!"

"Oh, the gash is there all right, partly covered by her hair."

He stood still, staring at me oddly.

"One meets with cases of singular devotion in unexpected quarters sometimes," he said.

"You mean that the woman inflicted the wound upon herself in order..."

'To save old Kwen Lung--exactly! It's marvelous."

"Good heavens!" I exclaimed. "And the window?"

"Oh! it was broken right enough--by two drunken sailormen fighting in the court outside! Sash and everything smashed to splinters."

He began irritably to pace the carpet again.

"It must have been a devil of a fight!" he added savagely.

"Meanwhile," said I, "where is old Kwen Lung hiding?"

"But more particularly," cried Harley, "where has he hidden the poor victim? Come along, Knox! I'm going down there for a final look round."

"Of course the premises are being watched?"

"Of course--and also, of course, I shall be the laughing stock of Scotland Yard if nothing results."

It was close on midnight when once more I found myself in Pennyfields. Carried away by Harley's irritable excitement I had quite forgotten the romance of Captain Dan; and when, having exchanged greetings with the detective on duty hard by the house of Kwen Lung, we presently found ourselves in the presence of Ma Lorenzo, I scarcely knew for a moment if I were "Jim" or my proper self.

"Is Kwen Lung in?" asked Harley sternly.

The woman shook her head.

"No," she replied; "he sometimes stop away a whole week."

"Does he?" jerked Harley. "Come in, Knox; we'll take another look round."

A moment later I found myself again in the room of the golden joss. The red curtain had been removed from before the shattered window, but otherwise the place looked exactly as it had looked before. The atmosphere was much less stale, however, but there was something repellent about the great gilded idol smiling eternally from his pedestal beside the door.

I stared into the leering face, and it was the face of one who knew and who might have said: "Yes! this and other things equally strange have I beheld in many lands as well as England. Much I could tell. Many things grim and terrible, and some few joyous; for behold! I smile but am silent."

For a while Harley stared abstractedly at the bloodstains on the pedestal of the joss and upon the floor beneath from which the matting had been pulled back. Suddenly he turned to Ma Lorenzo:

"Where have you hidden the body?" he demanded.

Watching her, I thought I saw the woman flinch, but there was enough of the Oriental in her composition to save her from self- betrayal. She shook her head slowly, watching Harley through half-closed eyes.

"Nobody hab," she replied.

And I thought for once that her lapse into pidgin had been deliberate and not accidental.

When finally we quitted the house of the missing Kwen Lung, and when, Harley having curtly acknowledged "good night" from the detective on duty, we came out into Limehouse Causeway.

"You have not overlooked the possibility, Harley," I said, "that this woman's explanation may be true, and that the fireman of the Seahawk may have been entertaining us with an account of a weird dream?"

"No!" snapped Harley--"neither will Scotland Yard overlook it."

He was in a particularly impossible mood, for he so rarely made mistakes that to be detected in one invariably brought out those petulant traits of character which may have been due in some measure to long residence in the East. Recognizing that he would rather be alone I parted from him at the corner of Chancery Lane and returned to my own chambers. Furthermore, I was very tired, for it was close upon two o'clock, and on turning in I very promptly went to sleep, nor did I awaken until late in the morning.

For some odd reason, but possibly because the fact had occurred to me just as I was retiring, I remembered at the moment of waking that I had not told Harley about the romantic wedding of Captain Dan. As I had left my friend in very ill humor I thought that this would be a good excuse for an early call, and just before eleven o'clock I walked into his office. Innes, his invaluable secretary, showed me into the study at the back.

"Hallo, Knox," said Harley, looking up from a little silver Buddha which he was examining, "have you come to ask for news of the Kwen Lung case?"

"No," I replied. "Is there any?"

Harley shook his head.

"It seems like fate," he declared, "that this thing should have been sent to me this morning."

He indicated the silver Buddha. "A present from a friend who knows my weakness for Chinese ornaments," he explained grimly. "It reminds me of that damned joss of Kwen Lung's!"

I took up the little image and examined it with interest. It was most beautifully fashioned in the patient Oriental way, and there was a little hinged door in the back which fitted so perfectly that when closed it was quite impossible to detect its presence. I glanced at Harley.

"I suppose you didn't find a jewel inside?" I said lightly.

"No," he replied; "there was nothing inside."

But even as he uttered the words his whole expression changed, and so suddenly as to startle me. He sprang up from the table, and:

"Have you an hour to spare, Knox?" he cried excitedly.

"I can spare an hour, but what for?"

"For Kwen Lung!"

Four minutes later we were speeding in the direction of Limehouse, and not a word of explanation to account for this sudden journey could I extract from my friend. Therefore I beguiled the time by telling him of my adventure with Captain Dan.

Harley listened to the story in unbroken silence, but at its termination he brought his hand down sharply on my knee.

"I have been almost perfectly blind, Knox," he said; "but not quite so perfectly blind as you!"

I stared at him in amazement, but he merely laughed and offered no explanation of his words.

Presently, then, I found myself yet again in the familiar room of the golden joss. Ma Lorenzo, in whom some hidden anxiety seemed to have increased since I had last seen her, stood at the top of the stairs watching us. Upon what idea my friend was operating and what he intended to do I could not imagine; but without a word to the woman he crossed the room and grasping the great golden idol with both arms he dragged it forward across the floor!

As he did so there was a stifled shriek, and Ma Lorenzo, stumbling down the steps, threw herself on her knees before Harley! Raising imploring hands:

"No, no!" she moaned. "Not until I tell you--I tell you everything first!"

"To begin with, tell me how to open this thing," he said sternly.

Momentarily she hesitated, and did not rise from her knees, but:

"Do you hear me?" he cried.

The woman rose unsteadily and walking slowly round the joss manipulated some hidden fastening, whereupon the entire back of the thing opened like a door! From what was within she shudderingly averted her face, but Harley, stepping back against the wall, stopped and peered into the cavity.

"Good God!" he muttered. "Come and look, Knox."

Prepared by his manner for some gruesome spectacle, I obeyed--and from that which I saw I recoiled in horror.

"Harley," I whispered, "Harley! who is it?"

The spectacle had truly sickened me. Crouched within the narrow space enclosed by the figure of the idol was the body of an old and wrinkled Chinaman! His knees were drawn up to his chin, and his head so compressed upon them that little of his features could be seen.

"It is Kwen Lung!" murmured Ma Lorenzo, standing with clasped hands and wild eyes over by the window. "Kwen Lung--and I am glad he is dead!"

Such a note of hatred came into her voice as I had never heard in the voice of any woman.

"He is vile, a demon, a mocking cruel demon! Long, long years ago I would have killed him, but always I was afraid. I tell you everything, everything. This is how he comes to be dead. The little one"--again her voice changed and a note of almost grotesque tenderness came into it--"the lotus-flower, that is his own daughter's child, flesh of his flesh, he keeps a prisoner as the women of China are kept, up there"--she raised one fat finger aloft--"up above. He does not know that someone comes to see her--someone who used to come to smoke but who gave it up because he had looked into the dear one's eye. He does not know that she goes with me to see her man. Ah! we think he does not know! I--I arrange it all. A week ago they were married. Tuesday night, when Kwen Lung die, I plan for her to steal away for ever, for ever."

Tears now were running down the woman's fat cheeks, and her voice quivered emotionally.

"For me it is the end, but for her it is the beginning of life. All right! I don't matter a damn! She is young and beautiful. Ah, God! so beautiful! A drunken pig comes here and finds his way in, so I give him the smoke and presently he sleeps, but it makes delay, and I don't know how soon Kwen Lung, that yellow demon, will wake. For he is like the bats who sleep all day and wake at night.

"At last the sailor pig sleeps and I call softly to my dear little one that the time has come. I have gone out into the street, locking the door behind me, to see if her man is waiting, and I hear her shrieks--her shrieks! I hurry back. My hands tremble so much that I can scarcely unlock the door. At last I enter, and I see and I know--that yellow devil has learned all and has been playing with us like cat and mouse! He is lashing her, with a great whip! Lashing her--that tiny, sweet flower. Ah!"

She choked in her utterance, and turning to the gilded joss which contained the dead Chinaman she shook her clenched hands at it, and the expression, on her face I can never forget. Then:

"As I shriek curses at him, crash goes the window--and I see her husband spring into the room! The tender one had fallen, there at the foot of the joss, and Kwen Lung, his teeth gleaming--like a rat--like a devill--turns to meet him. So he is when her man strike him, once. Just once, here." She rested her hand upon her heart. "And he falls--and he coughs. He lie still. For him it is finished. That devil heart has ceased to beat. Ah!"

She threw up her hands, and:

"That is all. I tell you no more."

"One thing more," said Harley sternly; "the name of the man who killed Kwen Lung?"

At that Ma Lorenzo slowly raised her head and folded her arms across her bosom. There was something one could never forget in the expression of her fat face.

"Not if you burn me alive!" she answered in a low voice. "No one ever knows that--from me."

She sank on to the divan and buried her face in her hands. Her fat shoulders shook grotesquely; and Harley stood perfectly still staring across at her for fully a minute. I could hear voices in the street outside and the hum of traffic in Limehouse Causeway.

Then my friend did a singular thing. Walking over to the gilded joss he reclosed the opening and not without a great effort pushed the great idol back against the wall.

"There are times, Knox," he said, staring at me oddly, "when I'm glad that I am not an official agent of the law."

While I watched him dumfounded he walked across to the woman and touched her on the shoulder. She raised her tear-stained face.

"All right," she whispered. "I am ready."

"Get ready as soon as you like," said he tersely.

"I'll have the man removed who is watching the house, and you can reckon on forty-eight hours to make yourself scarce."

With never another word he seized me by the arm and hurried me out of the place! Ten paces along the street a shabby-looking fellow was standing, leaning against a pillar. Harley stopped, and:

"Even the greatest men make mistakes sometimes, Hewitt," he remarked. "I'm throwing up the case; probably Inspector Wessex will do the same. Good morning."

On towards the Causeway he led me--for not a word was I capable of uttering; and just before we reached that artery of Chinatown, from down-river came the deep, sustained note of a steamer's siren, the warning of some big liner leaving dock.

"That will be the Patna," said Harley. "She sails at twelve o'clock, I think you said?"

MAN WITH THE SHAVEN SKULL

CHAPTER I

A STRANGE DISAPPEARANCE

"Pull that light lower," ordered Inspector Wessex. "There you are, Mr. Harley; what do you make of it?"

Paul Harley and I bent gingerly over the ghastly exhibit to which the C.I.D. official had drawn our attention, and to view which we had journeyed from Chancery Lane to Wapping.

This was the body of a man dressed solely in ragged shirt and trousers. But the remarkable feature of his appearance lay in the fact that every scrap of hair from chin, lip, eyebrows and skull had been shaved off!

There was another facial disfigurement, peculiarly and horribly Eastern, which my pen may not describe.

"Impossible to identify!" murmured Harley. "Yes, you were right, Inspector; this is a victim of Oriental deviltry. Look here, too!"

He indicated three small wounds, one situated on the left shoulder and the others on the forearm of the dead man.

"The divisional surgeon cannot account for them," replied Wessex. "They are quite superficial, and he thinks they may be due to the fact that the body got entangled with something in the river."

"They are due to the fact that the man had a birthmark on his shoulder and something-- probably a name or some device--tattooed on his arm," said Harley quietly. "Some few years ago, I met with a similar case in the neighborhood of Stambul. A woman," he added, significantly.

Detective-Inspector Wessex listened to my companion with respect, for apart from his established reputation as a private inquiry- agent which had made his name familiar in nearly every capital of the civilized world, Paul Harley's work in Constantinople during the six months preceding war with Turkey had merited higher reward than it had ever received. Had his recommendations been adopted the course of history must have been materially changed.

"You think it's a Chinatown case, then, Mr. Harley?"

"Possibly," was the guarded answer.

Paul Harley nodded to the constable in charge, and the ghastly figure was promptly covered up again. My friend stood staring vacantly at Wessex, and presently:

"The chief actor, I think, will prove to be not Chinese," he said, turned, and walked out.

"If there's any development," remarked Wessex as the three of us entered Harley's car, which stood at the door, "I will, of course, report to you, Mr. Harley. But in the absence of any clue or mark of identification, I fear the verdict will be, 'Body of a man unknown,' etc., which has marked the finish of a good many in this cheerful quarter of London."

"Quite so," said Harley, absently. "It presents extraordinary features, though, and may not end as you suppose. However--where do you want me to drop you, Wessex, at the Yard?"

"Oh no," answered Wessex. "I made a special visit to Wapping just to get your opinion on the shaven man. I'm really going down to Deepbrow to look into that new disappearance case; the daughter of the gamekeeper. You'll have read of it?"

"I have," said Harley shortly.

Indeed, readers of the daily press were growing tired of seeing on the contents bills: "Another girl missing." The circumstance (which might have been no more than coincidence) that three girls had disappeared within the last eight weeks leaving no trace behind, had stimulated the professional scribes to link the cases, although no visible link had been found, and to enliven a somewhat dull journalistic season with theories about "a new Mormon menace."

The vanishing of this fourth girl had inspired them to some startling headlines, and the case had interested me personally for the reason that I was acquainted with Sir Howard Hepwell, one of whose gamekeepers was the stepfather of the missing Molly Clayton. Moreover, it was hinted that she had gone away in the company of Captain Ronald Vane, at that time a guest of Sir Howard's at the Manor.

In fact, Sir Howard had 'phoned to ask me if I could induce Harley to run down, but my friend had expressed himself as disinterested in a common case of elopement. Now, as Wessex spoke, I glanced aside at Harley, wondering if the fact that so celebrated a member of the C.I.D. as Detective-Inspector Wessex had been put in charge would induce him to change his mind.

We were traversing a particularly noisy and unsavory section of the Commercial Road, and although I could see that Wessex was anxious to impart particulars of the case to Harley, so loud was the din that I recognized the impossibility of conversing, and therefore:

"Have you time to call at my rooms, Wessex?" I asked.

"Well," he replied, "I have three-quarters of an hour."

"You can do it in the car," said Harley suddenly. "I have been asked to look into this case myself, and before I definitely decline I should like to hear your version of the matter."

Accordingly, we three presently gathered in my chambers, and Wessex, with one eye on the clock, outlined the few facts at that time in his possession respecting the missing girl.

Two days before the news of the disappearance had been published broadcast under such headings as I have already indicated, a significant scene had been enacted in the gamekeeper's cottage.

Molly Clayton, a girl whose remarkable beauty had made her a central figure in numerous scandalous stories, for such is the charity of rural neighbors, was detected by her stepfather, about eight in the evening, slipping out of the cottage.

"Where be ye goin', hussy?" he demanded, grasping her promptly by the arm.

"For a walk!" she replied defiantly.

"A walk wi' that fine soger from t' Manor!" roared Bramber furiously. "You'll be sorry yet, you barefaced gadabout! Must I tell you again that t' man's a villain?"

The girl wrenched her arm from Bramber's grasp, and blazed defiance from her beautiful eyes.

"He knows how to respect a woman--what you don't!" she retorted hotly.

"So I don't respect you, my angel?" shouted her stepfather. "Then you know what you can do! The door's open and there's few'll miss you!"

Snatching her hat, the girl, very white, made to go out. Whereat the gamekeeper, a brutal man with small love for Molly, and maddened by her taking him at his word, seized her suddenly by her abundant fair hair and hauled her back into the room.

A violent scene followed, at the end of which Molly fainted and Bramber came out and locked the door.

When he came back about half-past nine the girl was missing. She did not reappear that night, and the police were advised in the morning. Their most significant discovery was this:

Captain Ronald Vane, on the night of Molly's disappearance, had left the Manor House, after dining alone with his host, Sir Howard Hepwell, saying that he proposed to take a stroll as far as the Deep Wood.

He never returned!

From the moment that Gamekeeper Bramber left his cottage, and the moment when Sir Howard Hepwell parted from his guest after dinner, the world to which these two people, Molly Clayton and Captain Vane, were known, knew them no more!

I was about to say that they were never seen again. But to me has fallen the task of relating how and where Paul Harley and I met with Captain Vane and Molly Clayton.

At the end of the Inspector's account:

"H'm," said Harley, glancing under his thick brows in my direction, "could you spare the time, Knox?"

"To go to Deepbrow?" I asked with interest.

"Yes; we have ten minutes to catch the train."

"I'll come," said I. "Sir Howard will be delighted to see you, Harley."

CHAPTER 2

THE CLUE OF THE PHOTOGRAPHS

"What do you make of it, Inspector?" asked my friend. Detective- Inspector Wessex smiled, and scratched his chin.

"There was no need for me to come down!" he replied. "And certainly no need for you, Mr. Harley!"

Harley bowed, smiling, at the implied compliment.

"It's a common or garden elopement!" continued the detective. "Vane's reputation is absolutely rotten, and the girl was clearly infatuated. He must have cared a good bit, too. He'll be cashiered, as sure as a gun!"

Leaving Sir Howard at the Manor, we had joined Inspector Wessex at a spot where the baronet's preserves bordered a narrow lane. Here the ground was soft, and the detective drew Harley's attention to a number of footprints by a stile.

"I've got evidence that he was seen here with the girl on other occasions. Now, Mr. Harley, I'll ask you to look over these footprints."

Harley dropped to his knees and made a brief but close examination of the ground round about. One particularly clear imprint of a pointed toe he noticed especially; and Wessex, diving into the pocket of his light overcoat, produced a patent- leather shoe, such as is used for evening wear.

"He had a spare pair in his bag," he explained nonchalantly, "and his man did not prove incorruptible!"

Harley took the shoe and placed it in the impression. It fitted perfectly!

"This is Molly Clayton, I take it?" he said, indicating the prints of a woman's foot.

"Yes," assented Wessex. "You'll notice that they stood for some little time and then walked off, very close together."

Harley nodded absently.

"We lose them along here," continued Wessex, leading up the lane; "but at the corner by the big haystack they join up with the tracks of a motor-car! I ask for nothing clearer! There was rain that afternoon, but there's been none since."

"What does the Captain's man think?"

"The same as I do! He's not surprised at any madness on Vane's part, with a pretty woman in the case!"

"The girl left nothing behind--no note?"

"Nothing."

"Traced the car?"

"No. It must have been hired or borrowed from a long distance off."

Where the tracks of the tires were visible we stopped, and Harley made a careful examination of the marks.

"Seems to have had a struggle with her," he said, dryly.

"Very likely!" agreed Wessex, without interest.

Harley crawled about on the ground for some time, to the great detriment of his Harris tweeds, but finally arose, a curious expression on his face--which, however, the detective evidently failed to observe.

We returned to the Manor House where Sir Howard was awaiting us, his good-humored red face more red than usual; and in the library, with its sporting prints and its works for the most part dealing with riding, hunting, racing, and golf (except for a sprinkling of Nat Gould's novels and some examples of the older workmanship of Whyte-Melville), we were presently comfortably ensconced. On a side table were placed a generous supply of liquid refreshments, cigars and cigarettes; so that we made ourselves quite comfortable, and Sir Howard restrained his indignation, until each had a glass before him and all were smoking.

"Now," he began, "what have you got to report, gentlemen? You, Inspector," he pointed with his cigar toward Wessex, "have seen Vane's man and all of you have been down to look at these damned tracks. I only want to hear one thing; that you expect to trace the disgraceful couple. I'll see to it"--his voice rose almost to a shout--"that Vane is kicked out of the service, and as to that shameless brat of Bramber's, I wish her no worse than the blackguard's company!"

"One moment, Sir Howard, one moment," said Harley quietly; "there are always two sides to a case."

"What do you mean, Mr. Harley? There's only one side that interests me--the outrage inflicted upon my hospitality by this dirty guest of mine. For the girl I don't give twopence; she was bound to come to a bad end."

"Well," said Harley, "before we pronounce the final verdict upon either of them I should like to interview Bramber. Perhaps," he added, turning to Wessex, "it would be as well if Mr. Knox and I went alone. The presence of an official detective sometimes awes this class of witness."

"Quite right, quite right!" agreed Sir Howard, waving his cigar vigorously. "Go and see Bramber, Mr. Harley; tell him that no blame attaches to himself whatever; also, tell him with my compliments that his stepdaughter is..."

"Quite so, quite so," interrupted Harley, endeavoring to hide a smile. "I understand your feelings, Sir Howard, but again I ask you to reserve your verdict until all the facts are before us."

As a result, Harley and I presently set out for the gamekeeper's cottage, and as the man had been warned that we should visit him, he was on the porch smoking his pipe. A big, dark, ugly fellow he proved to be, of a very forbidding cast of countenance. Having introduced ourselves:

"I always knowed she'd come to a bad end!" declared Gamekeeper Bramber, almost echoing Sir Howard's words. "One o' these gentlemen o' hers was sure to be the finish of her!"

"She had other admirers--before Captain Vane?"

"Aye! the hussy! There was a black-faced villain not six months since! He got t' vain cat to go to London an' have her photograph done in a dress any decent woman would 'a' blushed to look at! Like one o' these Venuses up at t' Manor! Good riddance! She took after her mother!"

The violent old ruffian was awkward to examine, but Harley persevered.

"This previous admirer caused her to be photographed in that way, did he? Have you a copy?"

"No!" blazed Bramber. "What I found I burnt! He ran off, like I told her he would--an' her cryin' her eyes out! But the pretty soger dried her tears quick enough!".

"Do you know this man's name?"

"No. A foreigner, he was."

"Where were the photographs done--in London, you say?"

"Aye."

"Do you know by what photographer?"

"I don't! An' I don't care! Piccadilly they had on 'em, which was good enough for me."

"Have you her picture?"

"No!"

"Did she receive a letter on the day of her disappearance?"

"Maybe."

"Good day!" said Harley. "And let me add that the atmosphere of her home was hardly conducive to ideal conduct!"

Leaving Bramber to digest this rebuke, we came out of the cottage. Dusk was falling now, and by the time that we regained the Manor the place was lighted up. Inspector Wessex was waiting for us in the library, and:

"Well?" he said, smiling slightly as we entered.

"Nothing much," replied Harley dryly, "except that I don't wonder at the girl's leaving such a home."

"What's that! What!" roared a big voice, and Sir Howard came into the room. "I tell you, Bramber only had one fault as a stepfather; he wasn't heavy-handed enough. A bad lot, sir, a bad lot!"

"Well, sir," said Inspector Wessex, looking from one to another, "personally, beyond the usual inquiries at railway stations, etc., I cannot see that we can do much here. Don't you agree with me, Mr. Harley?"

Harley nodded.

"Quite," he replied. "There is a late train to town which I think we could catch if we started at once."

"Eh?" roared Sir Howard; "you're not going back to-night? Your rooms are ready for you, damn it!"

"I quite appreciate the kindness, Sir Howard," replied Harley; "but I have urgent business to attend to in London. Believe me, my departure is unavoidable."

The blue eyes of the baronet gleamed with the simple cunning of his kind.

"You've got something up your sleeve," he roared. "I know you have, I know you have!"

Inspector Wessex looked at me significantly, but I could only shrug my shoulders in reply; for in these moods Harley was as inscrutable as the Sphinx.

However, he had his way, and Sir Howard hurriedly putting a car in commission, we raced for the local station and just succeeded in picking up the express at Claybury.

Wessex was rather silent throughout the journey, often glancing in my friend's direction, but Harley made no further reference to the case beyond outlining the interview with Bramber, until, as we were parting at the London terminus, Wessex to report to Scotland Yard and I to go to Harley's rooms:

"How long do you think it will take you to find that photographer, Wessex?" he asked.

"Piccadilly is a sufficient clue."

"Well," replied the Inspector, "nothing can be done to-night, of course, but I should think by mid-day tomorrow the matter should be settled."

"Right," said Harley shortly. "May I ask you to report the result to me, Wessex?"

"I will report without fail."

CHAPTER 3

ALI OF CAIRO

It was not until the evening of the following day that Harley rang me up, and:

"I want you to come round at once," he said urgently. "The Deepbrow case is developing along lines which I confess I had anticipated, but which are dramatic nevertheless."

Knowing that Harley did not lightly make such an assertion, I put aside the work upon which I was engaged and hurried around to Chancery Lane. I found my friend, pipe in mouth, walking up and down his smoke-laden study in a state which I knew to betoken suppressed excitement, and:

"Did Wessex find your photographer?" I asked on entering.

"Yes," he replied. "A first-class man, as I had anticipated. As I had further anticipated he did a number of copies of the picture for the foreign gentleman--about fifty, in fact!"

"Fifty!"

"Yes! Does the significance of that fact strike you?" asked Harley, a queer smile stealing across his tanned, clean-shaven face.

"It is an extraordinary thing for even an ardent admirer to have so many reproductions done of the same picture!"

"It is! I will show you now what I found trodden into one of the footprints where the struggle took place beside the car."

Harley produced a piece of thick silk twine.

"What is it?"

"It is a link, Knox--a link to seek which I really went down to Deepbrow." He stared at me quizzically, but my answering look must have been a blank one. "It is part of the tassel of one of those red cloth caps commonly called in England, a fez!"

He continued to stare at me and I to stare at the piece of silk; then:

"What is the next move?" I demanded. "Your new clue rather bewilders me."

"The next move," he said, "is to retire to the adjoining room and make ourselves look as much like a couple of Oriental commercial travelers as our correctly British appearance will allow!"

"What!" I cried.

"That's it!" laughed Harley. "I have a perpetual tan, and I think I can give you a temporary one which I keep in a bottle for the purpose."

Twenty minutes later, then, having quitted Harley's chambers by a back way opening into one of those old-world courts which abound in this part of the metropolis, two quietly attired Eastern gentlemen got into a cab at the corner of Chancery Lane and proceeded in the direction of Limehouse.

There are haunts in many parts of London whose very existence is unsuspected by all but the few; haunts unvisited by the tourist and even unknown to the copy-hunting pressman. Into a quiet thoroughfare not three minutes' walk from the busy life of West India Dock Road, Harley led the way. Before a door sandwiched in between the entrance to a Greek tobacconist's establishment and a boarded shop-front, he paused and turned to me.

"Whatever you see or hear," he cautioned, "express no surprise. Above all, show no curiosity."

He rang the bell beside the door, and almost immediately it was opened by a Negress, grossly and repellently ugly.

Harley pattered something in what sounded like Arabic, whereat the Negress displayed the utmost servility, ushering us into an ill-lighted passage with every evidence of respect. Following this passage to its termination, an inner door was opened, and a burst of discordant music greeted us, together with a wave of tobacco smoke. We entered.

Despite my friend's particular injunctions to the contrary I gave a start of amazement.

We stood in the doorway of a fairly large apartment having a divan round three of its sides. This divan was occupied by ten or a dozen men of mixed nationalities--Arabs, Greeks, lascars, and others. They smoked cigarettes for the most part and sipped Mokha from little cups. A girl was performing a wriggling dance upon the square carpet occupying the centre of the floor, accompanied by a Nubian boy who twanged upon a guitar, and by most of the assembled company, who clapped their hands to the music or droned a low, tuneless dirge.

Shortly after our entrance the performance terminated, and the girl retired through a curtained doorway at the farther end of the room. Our presence being now observed, suspicious glances were cast in our direction, and a very aged man, who sat smoking a narghli near the door by which the girl had made her exit, gravely waved towards us the amber mouthpiece which he held in his hand.

Harley walked straight across to him, I close at his heels. The light of a lamp which hung close by fell fully upon my friend's face; and, rising from his seat, the old man greeted him with the dignified and graceful salutation of the East. At his request we seated ourselves beside him, and, while we all three smoked excellent Turkish cigarettes, Harley and he conversed in a low tone. Suddenly, at some remark of my friend's, our strange host rose to his feet, an angry frown contracting his heavy eyebrows.

Silence fell upon the company.

In a loud and peremptory voice he called out something in Arabic.

Instantly I detected a fellow near the entrance door, and whom I had not hitherto observed, slipping furtively into the shadow, with a view, as I thought, to secret departure. He seemed to be deformed in some way and had the most evil, pock-marked face I had ever beheld in my life. Angrily, the majestic old man recalled him. Whereupon, with a sort of animal snarl quite indescribable, the fellow plucked out a knife! Two men who had been on the point of seizing him fell back, and:

"Hold him!" shouted Harley, springing forward--"hold him! It's Ali of Cairo!"

But Harley was too late. Turning, the strange and formidable- looking Oriental ran like the wind! Ere hand could be raised to stay him he was through the doorway!

"That settles it," said Harley grimly, as once more I found myself in a cab beside him. "I was right; but he'll forestall us!"

"Who will forestall us?" I asked in bewilderment.

"The biggest villain in Europe, Asia, or Africa!" cried my companion. "I have wasted precious time to-day. I might have known." He drummed irritably upon his knees. "The place we have just left is a sort of club, you understand, Knox, and Hakim is the proprietor or host as well as being an old gentleman of importance and authority in the Moslem world. I told him of my suspicions--which step I should have taken earlier--and they were instantly confirmed. My man was there--recognized me--and bolted! He'll forestall us."

"But my dear fellow," I said patiently--"who is this man, and what has he to do with the Deepbrow case?"

"He is the blackest scoundrel breathing!" answered Harley bitterly. "As to what he has to do with the case--why did he bolt? At any rate, I know where to find him now--and we may not be too late after all."

"But who and what is this man?"

"He is Ali of Cairo! As to what he is--you will soon learn."

CHAPTER 4

THE HOUSE BY THE RIVER

On quitting the singular Oriental club, Harley had first raced off to a public telephone, where he had spoken for some time--as I now divined--to Scotland Yard. For when we presently arrived at the headquarters of the Metropolitan Police, I was surprised to find Inspector Wessex awaiting us. Leaning out of the cab window:

"Yes?" called Harley excitedly. "Was I right?"

"You were, Mr. Harley," answered Wessex, who seemed to be no less excited than my companion. "I got the man's reply an hour ago."

"I knew it!" said Harley shortly. "Get in, Wessex; we haven't a minute to waste."

The Inspector joined us in the cab, having first given instructions to the chauffeur. As we set out once more:

"You have had very little time to make the necessary arrangements," continued my friend.

"Time enough," replied Wessex. "They will not be expecting us."

"I'm not so sure of it. One of the biggest villains in the civilized world recognized me three minutes before I called you up and then made good his escape. However, there is at least a fighting chance."

Little more was said from that moment until the end of the drive, both my companions seeming to be consumed by an intense eagerness to reach our destination. At last the cab drew up in a deserted street. I had rather lost my bearings; but I knew that we were once more somewhere in the Chinatown area, and:

"Follow us until we get into the house," Harley said to Inspector Wessex, "and wait out of sight. If you hear me blow this whistle, bring up the men you have posted--as quick as you like! But make it your particular business to see that no one gets out!"

Into a pitch-dark yard we turned, and I felt a shudder of apprehension upon observing that it was the entrance to a wharf. Dully gleaming in the moonlight, the Thames, that grave of many a ghastly secret, flowed beneath us. Emerging from the shadow of the archway, we paused before a door in the wall on our left.

At that moment something gleamed through the air, whizzed past my ear, and fell with a metallic jingle on the stones!

Instinctively we both looked up.

At an unlighted window on the first floor I caught a fleeting glimpse of a dark face.

"You were right!" I said. "Ali of Cairo has forestalled us!"

Harley stooped and picked up a knife with a broad and very curious blade. He slipped it into his pocket, nonchalantly.

"All evidence!" he said. "Keep in the shadow and bend down. I am going to stand on your shoulders and get into that window!"

Wondering at his daring, I nevertheless obeyed; and Harley succeeded, although not without difficulty, in achieving his purpose. A moment after he had disappeared in the blackness of the room above.

"Stand clear, Knox!" I heard.

Two of the cushion seats sometimes called "poof-ottomans" were thrown down, and:

"Up you come!" called Harley. "I'll grasp your hands if you can reach."

It proved no easy task, but I finally managed to scramble up beside my friend--to find myself in a dark and stuffy little room.

"This way!" said Harley rapidly--"upstairs."

He led the way without more ado, but it was with serious misgivings that I stumbled up a darkened stair in the rear of my greatly daring friend.

A pistol cracked in the darkness--and my fez was no longer on my head!

Harley's repeater answered, and we stumbled through a heavily curtained door into a heated room, the air of which was laden with some Eastern perfume. In the dim light from a silken-shaded lantern a figure showed, momentarily, darting across the place before us.

Again Harley's pistol spoke, but, as it seemed, ineffectively.

I had little enough opportunity to survey my surroundings; yet even in those brief, breathless moments I saw enough of the place wherein we stood to make me doubt the evidence of my senses! Outside, I knew, lay a dingy wharf, amid a maze of mean streets; here was an opulently furnished apartment with a strong Oriental note in the decorations!

Snatching an electric torch from his pocket, Harley leaped through a doorway draped with rich Persian tapestry, and I came close on his heels. Outside was darkness. A strong draught met us; and, passing along a carpeted corridor, we never halted until we came to a room filled with the weirdest odds and ends, apparently collected from every quarter of the globe.

Crack!

A bullet flattened itself on the wall behind us!

"Good job he can't shoot straight!" rapped Harley.

The ray of the torch suddenly picked out the head and shoulders of a man who was descending through a trap in the floor! Ere we had time to shoot he was gone! I saw his brown fingers relax their hold--and a bundle which he had evidently hoped to take with him was left lying upon the floor.

Together we ran to the trap and looked down.

Slowly moving tidal water flowed darkly beneath us! For twenty breathless seconds we watched--but nothing showed upon the surface.

"I hope his swimming is no better than his shooting," I said.

"It can avail him little," replied Harley grimly; "a river-police boat is waiting for anyone who tries to escape from that side of the house. We are by no means alone in this affair, Knox. But, firstly, what have we here!" He took up the bundle which the fugitive had deserted. "Something incriminating when Ali of Cairo dared not stay to face it out! He would never have deserted this place in the ordinary way. That fellow who was such a bad shot was left behind, when the news of our approach reached here, to make a desperate attempt to remove some piece of evidence! I'll swear to it. But we were too soon for him!"

All the time he was busily removing the pieces of sacking and scraps of Oriental stuff with which the bundle was fastened; and finally he drew out a dress-suit, together with the linen, collar, shoes, and underwear--a complete outfit, in fact--and on top of the whole was a soft gray felt hat!

Eagerly Harley searched the garments for some name of a maker by which their owner might be identified. Presently, inside the lining of the breast pocket, where such a mark is usually found, he discovered the label of a well-known West End firm.

"The police can confirm it, Knox!" he said, looking up, his face slightly flushed with triumph; "but I, personally, have no doubt!"

"You may have no doubt, Harley," I retorted, "but I am full of doubt! What is the significance of this discovery to which you seem to attach so much importance?"

"At the moment," replied my friend, "never mind; I still have hopes--although they have grown somewhat slender--of making a much more important discovery."

"Why not permit the police to aid in the search?"

"The police are more useful in their present occupation," he replied. "We are dealing with the most cunning knave produced by East or West, and I don't mean to let him slip through my fingers if he is in this house! Nevertheless, Knox, I am submitting you to rather an appalling risk, I know; for our man is desperate, and if he is still in the place will prove as dangerous as a cornered rat."

"But the man who dropped through the trap?"

"The man who dropped through the trap," said Harley, "was not Ali of Cairo--and it is Ali of Cairo for whom I am looking!"

"The hunchback we saw to-night?"

Harley nodded, and having listened intently for a few moments, proceeded again to search the singular apartments of the abode. In each was evidence of Oriental occupancy; indeed, some of the rooms possessed a sort of Arabian Nights atmosphere. But no living creature was to be seen or heard anywhere. It was while the two of us, having examined every inch of wall, I should think, in the building, were standing staring rather blankly at each other in the room with the lighted lantern, that I saw Harley's expression change.

"Why," he muttered, "is this one room illuminated--and all the others in darkness?"

Even then the significance of this circumstance was not apparent to me. But Harley stared critically at an electric switch which was placed on the immediate right of the door and then up at the silk-shaded lantern which lighted the room. Crossing, he raised and lowered the switch rapidly, but the lamp continued to burn uninterruptedly!

"Ah!" he said--"a good trick!"

Grasping the wooden block to which the switch was attached, he turned it bodily--and I saw that it was a masked knob; for in the next moment he had pulled open the narrow section of wall--which proved to be nothing less than a cunningly fitted door!

A small, dimly lighted apartment was revealed, the Oriental note still predominant in its appointments, which, however, were few, and which I scarcely paused to note. For lying upon a mattress in this place was a pretty, fair-haired girl!

She lay on her side, having one white arm thrown out and resting limply on the floor, and she seemed to be in a semi-conscious condition, for although her fine eyes were widely opened, they had a glassy, witless look, and she was evidently unaware of our presence.

"Look at her pupils," rapped Harley. "They have drugged her with bhang! Poor, pretty fool!"

"Good God!" I cried. "Who is this, Harley?"

"Molly Clayton!" he answered. "Thank heaven we have saved one victim from Ali of Cairo."

CHAPTER 5

THE HAREM AGENCY

Owing to the instrumentality of Paul Harley, the public never learned that the awful riverside murder called by the Press in reference to the victim's shaven skull "the barber atrocity" had any relation to the Deepbrow case. It was physically impossible to identify the victim, and Harley had his own reasons for concealing the truth. The house on the wharf with its choice Oriental furniture was seized by the police; but, strange to relate, no arrest was made in connection with this most gruesome outrage. The man who dropped through the trap had been wounded by one of Harley's shots, and he sank for the last time under the very eyes of the crew of the police cutter.

It was at a late hour on the night of this concluding tragedy that I learned the amazing truth underlying the case. Wessex was still at work in the East End upon the hundred and one formalities which attached to his office, and Harley and I sat in the study of my friend's chambers in Chancery Lane.

"You see," Harley was explaining. "I got my first clue down at Deepbrow. The tracks leading to the motor-car. They showed--to anyone not hampered by a preconceived opinion--that the girl and Vane had not gone on together (since the man's footprints proved him to have been running), but that she had gone first and that he had run after her! Arguments: (a) He heard the approach of the car; or (b) he heard her call for help. In fact, it almost immediately became evident to me that someone else had met her at the end of the lane; probably someone who expected her, and whom she was going to meet when she, accidentally, encountered Vane! The captain was not attired for an elopement, and, more significant still, he said he should stroll to the Deep Wood, and that was where he did stroll to; for it borders the road at this point!

"I had privately ascertained, from the postman, that Molly Clayton actually received a letter on that morning! This resolved my last doubt. She was not going to meet Vane on the night of her disappearance.

"Then whom?"

"The old love! He who some months earlier had had over fifty seductive pictures of this undoubtedly pretty girl prepared for a purpose of his own!"

"Vane interfered?"

"When the girl saw that they meant to take her away, she no doubt made a fuss! He ran to the rescue! They had not reckoned on his being there, but these are clever villains, who leave no clues-- except for one who has met them on their own ground!"

"On their own ground! What do you mean, Harley? Who are these people?"

"Well--where do you suppose those fifty photographs went?"

"I cannot conjecture!"

"Then I will tell you. The turmoil in the East has put wealth and power into unscrupulous hands. But even before the war there were marts, Knox--open marts--at which a Negro girl might be purchased for some 30 pounds, and a Circassian for anything from 250 pounds to 500 pounds! Ah! You stare! But I assure you it was so. Here is the point, though: there were, and still are, private dealers! Those photographs were circulated among the nouveaux riches of the East! They were employed in the same way that any other merchant employs a catalogue. They reached the hands of many an opulent and abandoned 'profiteer' of Damascus, Stambul--where you will. Molly's picture would be one of many. Remember that hundreds of pretty girls disappear from their homes--taking the whole of the world--every year. Clearly, English beauty is popular at the moment! And," he added bitterly, "the arch-villain has escaped!"

"Ali of Cairo!" I cried. "Then Ali of Cairo..."

"Is the biggest slave-dealer in the East!"

"Good God! Harley--at last I understand!"

"I was slow enough to understand it myself, Knox. But once the theory presented itself I asked Wessex to get into immediate touch with the valet he had already interviewed at Deepbrow. It was the result of his inquiry to which he referred when we met him at Scotland Yard to-night. Captain Vane had a large mole on his shoulder and a girl's name, together with a small device, tattooed on his forearm--a freak of his Sandhurst days..."

"Then 'the man with the shaven skull'..."

"Is Captain Ronald Vane! May he rest in peace. But I never shall until the crook-back dealer in humanity has met his just deserts."

THE WHITE HAT

CHAPTER 1

MAJOR JACK RAGSTAFF

"Hallo! Innes," said Paul Harley as his secretary entered. "Someone is making a devil of a row outside."

"This is the offender, Mr. Harley," said Innes, and handed my friend a visiting card.

Glancing at the card, Harley read aloud:

"Major J. E. P. Ragstaff, Cavalry Club."

Meanwhile a loud harsh voice, which would have been audible in a full gale, was roaring in the lobby.

"Nonsense!" I could hear the Major shouting. "Balderdash! There's more fuss than if I had asked for an interview with the Prime Minister. Piffle! Balderdash!"

Innes's smile developed into a laugh, in which Harley joined, then:

"Admit the Major," he said.

Into the study where Harley and I had been seated quietly smoking, there presently strode a very choleric Anglo-Indian. He wore a horsy check suit and white spats, and his tie closely resembled a stock. In his hand he carried a heavy malacca cane, gloves, and one of those tall, light-gray hats commonly termed white. He was below medium height, slim and wiry; his gait and the shape of his legs, his build, all proclaimed the dragoon. His complexion was purple, and the large white teeth visible beneath a bristling gray moustache added to the natural ferocity of his appearance. Standing just within the doorway:

"Mr. Paul Harley?" he shouted.

It was apparently an inquiry, but it sounded like a reprimand.

My friend, standing before the fireplace, his hands in his pockets and his pipe in his mouth, nodded brusquely.

"I am Paul Harley," he said. "Won't you sit down?"

Major Ragstaff, glancing angrily at Innes as the latter left the study, tossed his stick and gloves on to a settee, and drawing up a chair seated himself stiffly upon it as though he were in a saddle. He stared straight at Harley, and:

"You are not the sort of person I expected, sir," he declared. "May I ask if it is your custom to keep clients dancin' on the mat and all that--on the blasted mat, sir?"

Harley suppressed a smile, and I hastily reached for my cigarette-case which I had placed upon the mantelshelf.

"I am always naturally pleased to see clients, Major Ragstaff," said Harley, "but a certain amount of routine is necessary even in civilian life. You had not advised me of your visit, and it is contrary to my custom to discuss business after five o'clock."

As Harley spoke the Major glared at him continuously, and then:

"I've seen you in India!" he roared; "damme! I've seen you in India!--and, yes! in Turkey! Ha! I've got you now sir!" He sprang to his feet. "You're the Harley who was in Constantinople in 1912."

"Quite true."

"Then I've come to the wrong shop."

"That remains to be seen, Major."

"But I was told you were a private detective, and all that."

"So I am," said Harley quietly. "In 1912 the Foreign Office was my client. I am now at the service of anyone who cares to employ me."

"Hell!" said the Major.

He seemed to be temporarily stricken speechless by the discovery that a man who had acted for the British Government should be capable of stooping to the work of a private inquiry agent. Staring all about the room with a sort of naive wonderment, he drew out a big silk handkerchief and loudly blew his nose, all the time eyeing Harley questioningly. Replacing his handkerchief he directed his regard upon me, and:

"This is my friend, Mr. Knox," said Harley; "you may state your case before him without hesitation, unless..."

I rose to depart, but:

"Sit down, Mr. Knox! Sit down, sir!" shouted the Major. "I have no dirty linen to wash, no skeletons in the cupboard or piffle of that kind. I simply want something explained which I am too thick-headed--too damned thick-headed, sir--to explain myself."

He resumed his seat, and taking out his wallet extracted from it a small newspaper cutting which he offered to Harley.

"Read that, Mr. Harley," he directed. "Read it aloud."

Harley read as follows:

"Before Mr. Smith, at Marlborough Street Police Court, John Edward Bampton was charged with assaulting a well-known clubman in Bond Street on Wednesday evening. It was proved by the constable who made the arrest that robbery had not been the motive of the assault, and Bampton confessed that he bore no grudge against the assailed man, indeed, that he had never seen him before. He pleaded intoxication, and the police surgeon testified that although not actually intoxicated, his breath had smelled strongly of liquor at the time of his arrest. Bampton's employers testified to a hitherto blameless character, and as the charge was not pressed the man was dismissed with a caution."

Having read the paragraph, Harley glanced at the Major with a puzzled expression.

"The point of this quite escapes me," he confessed.

"Is that so?" said Major Ragstaff. "Is that so, sir? Perhaps you will be good enough to read this."

From his wallet he took a second newspaper cutting, smaller than the first, and gummed to a sheet of club notepaper. Harley took it and read as follows:

"Mr. De Lana, a well-known member of the Stock Exchange, who met with a serious accident recently, is still in a precarious condition."

The puzzled look on Harley's face grew more acute, and the Major watched him with an expression which I can only describe as one of fierce enjoyment.

"You're thinkin' I'm a damned old fool, ain't you?" he shouted suddenly.

"Scarcely that," said Harley, smiling slightly, "but the significance of these paragraphs is not apparent, I must confess. The man Bampton would not appear to be an interesting character, and since no great damage has been done, his drunken frolic hardly comes within my sphere. Of Mr. De Lana, of the Stock Exchange, I never heard, unless he happens to be a member of the firm of De Lana and Day?"

"He's not a member of that firm, sir," shouted the Major. "He was, up to six o'clock this evenin'."

"What do you mean exactly?" inquired Harley, and the tone of his voice suggested that he was beginning to entertain doubts of the Major's sanity or sobriety; then:

"He's dead!" declared the latter. "Dead as the Begum of Bangalore! He died at six o'clock. I've just spoken to his widow on the telephone."

I suppose I must have been staring very hard at the speaker, and certainly Harley was doing so, for suddenly directing his fierce gaze toward me:

"You're completely treed, sir, and so's your friend!" shouted Major Ragstaff.

"I confess it," replied Harley quietly; "and since my time is of some little value I would suggest, without disrespect, that you explain the connection, if any, between yourself, the drunken Bampton, and Mr. De Lana, of the Stock Exchange, who died, you inform us, at six o'clock this evening as the result, presumably, of injuries received in an accident."

"That's what I'm here for!" cried Major Ragstaff. "In the first place, then, I am the party, although I saw to it that my name was kept out of print, whom the drunken lunatic assaulted."

Harley, pipe in hand, stared at the speaker perplexedly.

"Understand me," continued the Major, "I am the person--I, Jack Ragstaff--he assaulted. I was walkin' down from my quarters in Maddox Street on my way to dine at the club, same as I do every night o' my life, when this flamin' idiot sprang upon me, grabbed my hat"--he took up his white hat to illustrate what had occurred--"not this one, but one like it--pitched it on the ground and jumped on it!"

Harley was quite unable to conceal his smiles as the excited old soldier dropped his conspicuous head-gear on the floor and indulged in a vigorous pantomime designed to illustrate his statement.

"Most extraordinary," said Harley. "What did you do?"

"What did I do?" roared the Major. "I gave him a crack on the head with my cane, and I said things to him which couldn't be repeated in court. I punched him, and likewise hoofed him, but the hat was completely done in. Damn crowd collected, hearin' me swearin' and bellowin'. Police and all that; names an' addresses and all that balderdash. Man lugged away to guard-room and me turnin' up at the club with no hat. Damn ridiculous spectacle at my time of life."

"Quite so," said Harley soothingly; "I appreciate your annoyance, but I am utterly at a loss to understand why you have come here, and what all this has to do with Mr. De Lana, of the Stock Exchange."

"He fell out of the window!" shouted the Major.

"Fell out of a window?"

"Out of a window, sir, a second floor window ten yards up a side street! Pitched on his skull--marvel he wasn't killed outright!"

A faint expression of interest began to creep into Harley's glance, and:

"I understand you to mean, Major Ragstaff," he said deliberately, "that while your struggle with the drunken man was in progress Mr. De Lana fell out of a neighboring window into the street?"

"Right!" shouted the Major. "Right, sir!"

"Do you know this Mr. De Lana?"

"Never heard of him in my life until the accident occurred. Seems to me the poor devil leaned out to see the fun and overbalanced. Felt responsible, only natural, and made inquiries. He died at six o'clock this evenin', sir."

"H'm," said Harley reflectively. "I still fail to see where I come in. From what window did he fall?"

"Window above a sort of teashop, called Cafe Dame--damn silly name. Place on a corner. Don't know name of side street."

"H'm. You don't think he was pushed out, for instance?"

"Certainly not!" shouted the Major; "he just fell out, but the point is, he's dead!"

"My dear sir," said Harley patiently, "I don't dispute that point; but what on earth do you want of me?"

"I don't know what I want!" roared the Major, beginning to walk up and down the room, "but I know I ain't satisfied, not easy in my mind, sir. I wake up of a night hearin' the poor devil's yell as he crashed on the pavement. That's all wrong. I've heard hundreds of death-yells, but"--he took up his malacca cane and beat it loudly on the table--"I haven't woke up of a night dreamin' I heard 'em again."

"In a word, you suspect foul play?"

"I don't suspect anything!" cried the other excitedly, "but someone mentioned your name to me at the club--said you could see through concrete, and all that--and here I am. There's something wrong, radically wrong. Find out what it is and send the bill to me. Then perhaps I'll be able to sleep in peace."

He paused, and again taking out the large silk handkerchief blew his nose loudly. Harley glanced at me in rather an odd way, and then:

"There will be no bill, Major Ragstaff," he said; "but if I can see any possible line of inquiry I will pursue it and report the result to you."

CHAPTER 2

A CURIOUS OUTRAGE

"What do you make of it, Harley?" I asked. Paul Harley returned a work of reference to its shelf and stood staring absently across the study.

"Our late visitor's history does not help us much," he replied. "A somewhat distinguished army career, and so forth, and his only daughter, Sybil Margaret, married the fifth Marquis of Ireton. She is, therefore, the noted society beauty, the Marchioness of Ireton. Does this suggest anything to your mind?"

"Nothing whatever," I said blankly.

"Nor to mine," murmured Harley.

The telephone bell rang.

"Hallo!" called Harley. "Yes. That you, Wessex? Have you got the address? Good. No, I shall remember it. Many thanks. Good-bye."

He turned to me.

"I suggest, Knox," he said, "that we make our call and then proceed to dinner as arranged."

Since I was always glad of an opportunity of studying my friend's methods I immediately agreed, and ere long, leaving the lights of the two big hotels behind, our cab was gliding down the long slope which leads to Waterloo Station. Thence through crowded, slummish high-roads we made our way via Lambeth to that dismal thoroughfare, Westminster Bridge Road, with its forbidding, often windowless, houses, and its peculiar air of desolation.

The house for which we were bound was situated at no great distance from Kensington Park, and telling the cabman to wait, Harley and I walked up a narrow, paved path, mounted a flight of steps, and rang the bell beside a somewhat time-worn door, above which was an old-fashioned fanlight dimly illuminated from within.

A considerable interval elapsed before the door was opened by a marvelously untidy servant girl who had apparently been interrupted in the act of black-leading her face. Partly opening the door, she stared at us agape, pushing back wisps of hair from her eyes and with every movement daubing more of some mysterious black substance upon her countenance.

"Is Mr. Bampton in?" asked Harley.

"Yus, just come in. I'm cookin' his supper."

"Tell him that two friends of his have called on rather important business."

"All right," said the black-faced one. "What name is it?"

"No name. Just say two friends of his."

Treating us to a long, vacant stare and leaving us standing on the step, the maid (in whose hand I perceived a greasy fork) shuffled along the passage and began to mount the stairs. An unmistakable odor of frying sausages now reached my nostrils. Harley glanced at me quizzically, but said nothing until the Cinderella came stumbling downstairs again. Without returning to where we stood:

"Go up," she directed. "Second floor, front. Shut the door, one of yer."

She disappeared into gloomy depths below as Harley and I, closing the door behind us, proceeded to avail ourselves of the invitation. There was very little light on the staircase, but we managed to find our way to a poorly furnished bed-sitting-room where a small table was spread for a meal. Beside the table, in a chintz-covered arm-chair, a thick-set young man was seated smoking a cigarette and having a copy of the Daily Telegraph upon his knees.

He was a very typical lower middle-class, nothing-in-particular young man, but there was a certain truculence indicated by his square jaw, and that sort of self-possession which sometimes accompanies physical strength was evidenced in his manner as, tossing the paper aside, he stood up.

"Good evening, Mr. Bampton," said Harley genially. "I take it"-- pointing to the newspaper--"that you are looking for a new job?"

Bampton stared, a suspicion of anger in his eyes, then, meeting the amused glance of my friend, he broke into a smile very pleasing and humorous. He was a fresh-colored young fellow with hair inclined to redness, and smiling he looked very boyish indeed.

"I have no idea who you are," he said, speaking with a faint north-country accent, "but you evidently know who I am and what has happened to me."

"Got the boot?" asked Harley confidentially.

Bampton, tossing the end of his cigarette into the grate, nodded grimly.

"You haven't told me your name," he said, "but I think I can tell you your business." He ceased smiling. "Now look here, I don't want any more publicity. If you think you are going to make a funny newspaper story out of me change your mind as quick as you like. I'll never get another job in London as it is. If you drag me any further into the limelight I'll never get another job in England."

"My dear fellow," replied Harley soothingly, at the same time extending his cigarette-case, "you misapprehend the object of my call. I am not a reporter."

"What!" said Bampton, pausing in the act of taking a cigarette, "then what the devil are you?"

"My name is Paul Harley, and I am a criminal investigator."

He spoke the words deliberately, having his eyes fixed upon the other's face; but although Bampton was palpably startled there was no trace of fear in his straightforward glance. He took a cigarette from the case, and:

"Thanks, Mr. Harley," he said. "I cannot imagine what business has brought you here."

"I have come to ask you two questions," was the reply. "Number one: Who paid you to smash Major Ragstaff's white hat? Number two: How much did he pay you?"

To these questions I listened in amazement, and my amazement was evidently shared by Bampton. He had been in the act of lighting his cigarette, but he allowed the match to burn down nearly to his fingers and then dropped it with a muttered exclamation in the fire. Finally:

"I don't know how you found out," he said, "but you evidently know the truth. Provided you assure me that you are not out to make a silly-season newspaper story, I'll tell you all I know."

Harley laid his card on the table, and:

"Unless the ends of justice demand it," he said, "I give you my word that anything you care to say will go no further. You may speak freely before my friend, Mr. Knox. Simply tell me in as few words as possible what led you to court arrest in that manner."

"Right," replied Bampton, "I will." He half closed his eyes, reflectively. "I was having tea in the Lyons' cafe, to which I always go, last Monday afternoon about four o'clock, when a man sat down facing me and got into conversation."

"Describe him!"

"He was a man rather above medium height. I should say about my own build; dark, going gray. He had a neat moustache and a short beard, and the look of a man who had traveled a lot. His skin was very tanned, almost as deeply as yours, Mr. Harley. Not at all the sort of chap that goes in there as a rule. After a while he made an extraordinary proposal. At first I thought he was joking, then when I grasped the idea that he was serious I concluded he was mad. He asked me how much a year I earned, and I told him Peters and Peters paid me 150 pounds. He said: 'I'll give you a year's salary to knock a man's hat off!'"

As Bampton spoke the words he glanced at us with twinkling eyes, but although for my own part I was merely amused, Harley's expression had grown very stern.

"Of course, I laughed," continued Bampton, "but when the man drew out a fat wallet and counted ten five-pound notes on the table I began to think seriously about his proposal. Even supposing he was cracked, it was absolutely money for nothing.

"'Of course,' he said, 'you'll lose your job and you may be arrested, but you'll say that you had been out with a few friends and were a little excited, also that you never could stand white hats. Stick to that story and the balance of a hundred pounds will reach you on the following morning.'

"I asked him for further particulars, and I asked him why he had picked me for the job. He replied that he had been looking for some time for the right man; a man who was strong enough physically to accomplish the thing, and someone"--Bampton's eyes twinkled again--"with a dash of the devil in him, but at the same time a man who could be relied upon to stick to his guns and not to give the game away.

"You asked me to be brief, and I'll try to be. The man in the white hat was described to me, and the exact time and place of the meeting. I just had to grab his white hat, smash it, and face the music. I agreed. I don't deny that I had a couple of stiff drinks before I set out, but the memory of that fifty pounds locked up here in my room and the further hundred promised, bucked me up wonderfully. It was impossible to mistake my man; I could see him coming toward me as I waited just outside a sort of little restaurant called the Cafe Dame. As arranged, I bumped into him, grabbed his hat and jumped on it."

He paused, raising his hand to his head reminiscently.

"My man was a bit of a scrapper," he continued, "and he played hell. I've never heard such language in my life, and the way he laid about me with his cane is something I am not likely to forget in a hurry. A crowd gathered, naturally, and (also naturally) I was 'pinched.' That didn't matter much. I got off lightly; and although I've been dismissed by Peters and Peters, twenty crisp fivers are locked in my trunk there, with the ten which I received in the City."

Harley checked him, and:

"May I see the envelope in which they arrived?" he asked.

"Sorry," replied Bampton, "but I burned it. I thought it was playing the game to do so. It wouldn't have helped you much, though," he added; "It was an ordinary common envelope, posted in the City, address typewritten, and not a line enclosed."

"Registered?"

"No."

Bampton stood looking at us with a curious expression on his face, and suddenly:

"There's one point," he said, "on which my conscience isn't easy. You know about that poor devil who fell out of a window? Well, it would never have happened if I hadn't kicked up a row in the street. There's no doubt he was leaning out to see what the disturbance was about when the accident occurred."

"Did you actually see him fall?" asked Harley.

"No. He fell from a window several yards behind me in the side street, but I heard him cry out, and as I was lugged off by the police I heard the bell of the ambulance which came to fetch him."

He paused again and stood rubbing his head ruefully.

"H'm," said Harley; "was there anything particularly remarkable about this man in the Lyons' cafe?"

Bampton reflected silently for some moments, and then:

"Nothing much," he confessed. "He was evidently a gentleman, wore a blue top-coat, a dark tweed suit, and what looked like a regimental tie, but I didn't see much of the colors. He was very tanned, as I have said, even to the backs of his hands--and oh, yes! there was one point: He had a gold-covered tooth."

"Which tooth?"

"I can't remember, except that it was on the left side, and I always noticed it when he smiled."

"Did he wear any ring or pin which you would recognize?"

"No."

"Had he any oddity of speech or voice?"

"No. Just a heavy, drawling manner. He spoke like thousands of other cultured Englishmen. But wait a minute--yes! There was one other point. Now I come to think of it, his eyes very slightly slanted upward."

Harley stared.

"Like a Chinaman's?"

"Oh, nothing so marked as that. But the same sort of formation."

Harley nodded briskly and buttoned up his overcoat.

"Thanks, Mr. Bampton," he said; "we will detain you no longer!"

As we descended the stairs, where the smell of frying sausages had given place to that of something burning--probably the sausages:

"I was half inclined to think that Major Ragstaff's ideas were traceable to a former touch of the sun," said Harley. "I begin to believe that he has put us on the track of a highly unusual crime. I am sorry to delay dinner, Knox, but I propose to call at the Cafe Dame."

CHAPTER 3

A CRIMINAL GENIUS

On entering the doorway of the Cafe Dame we found ourselves in a narrow passage. In front of us was a carpeted stair, and to the right a glass-paneled door communicating with a discreetly lighted little dining room which seemed to be well patronized. Opening the door Harley beckoned to a waiter, and:

"I wish to see the proprietor," he said.

114

"Mr. Meyer is engaged at the moment, sir," was the reply.

"Where is he?"

"In his office upstairs, sir. He will be down in a moment."

The waiter hurried away, and Harley stood glancing up the stairs as if in doubt what to do.

"I cannot imagine how such a place can pay," he muttered. "The rent must be enormous in this district."

But even before he ceased speaking I became aware of an excited conversation which was taking place in some apartment above.

"It's scandalous!" I heard, in a woman's shrill voice. "You have no right to keep it! It's not your property, and I'm here to demand that you give it up."

A man's voice replied in voluble broken English, but I could only distinguish a word here and there. I saw that Harley was interested, for catching my questioning glance, he raised his finger to his lips enjoining me to be silent.

"Oh, that's the game, is it?" continued the female voice. "Of course you know it's blackmail?"

A flow of unintelligible words answered this speech, then:

"I shall come back with someone," cried the invisible woman, "who will make you give it up!"

"Knox," whispered Harley in my ear, "when that woman comes down, follow her! I'm afraid you will bungle the business, and I would not ask you to attempt it if big things were not at stake. Return here; I shall wait."

As a matter of fact, his sudden request had positively astounded me, but ere I had time for any reply a door suddenly banged open above and a respectable-looking woman, who might have been some kind of upper servant, came quickly down the stairs. An expression of intense indignation rested upon her face, and without seeming to notice our presence she brushed past us and went out into the street.

"Off you go, Knox!" said Harley.

Seeing myself committed to an unpleasant business, I slipped out of the doorway and detected the woman five or six yards away hurrying in the direction of Piccadilly. I had no difficulty in following her, for she was evidently unsuspicious of my presence, and when presently she mounted a westward-bound 'bus I did likewise, but while she got inside I went on top, and occupied a seat on the near side whence I could observe anyone leaving the vehicle.

If I had not known Paul Harley so well I should have counted the whole business a ridiculous farce, but recognizing that something underlay these seemingly trivial and disconnected episodes, I lighted a cigarette and resigned myself to circumstance.

At Hyde Park Corner I saw the woman descending, and when presently she walked up Hamilton Place I was not far behind her. At the door of an imposing mansion she stopped, and in response to a ring of the bell the door was opened by a footman, and the woman hurried in. Evidently she was an inmate of the establishment; and conceiving that my duty was done when I had noted the number of the house, I retraced my steps to the corner; and, hailing a taxicab, returned to the Cafe Dame.

On inquiring of the same waiter whom Harley had accosted whether my friend was there:

"I think a gentleman is upstairs with Mr. Meyer," said the man.

"In his office?"

"Yes, sir."

Thereupon I mounted the stairs and before a half-open door paused. Harley's voice was audible within, and therefore I knocked and entered.

I discovered Harley standing by an American desk. Beside him in a revolving chair which, with the desk, constituted the principal furniture of a tiny office, sat a man in a dress-suit which had palpably not been made for him. He had a sullen and suspiciously Teutonic cast of countenance, and he was engaged in a voluble but hardly intelligible speech as I entered.

"Ha, Knox!" said Harley, glancing over his shoulder, "did you manage?"

"Yes," I replied.

Harley nodded shortly and turned again to the man in the chair.

"I am sorry to give you so much trouble, Mr. Meyer," he said, "but I should like my friend here to see the room above."

At this moment my attention was attracted by a singular object which lay upon the desk amongst a litter of bills and accounts. This was a piece of rusty iron bar somewhat less than three feet in length, and which once had been painted green.

"You are looking at this tragic fragment, Knox," said Harley, taking up the bar. "Of course"--he shrugged his shoulders--"it explains the whole unfortunate occurrence. You see there was a flaw in the metal at this end, here"--he indicated the spot--"and the other end had evidently worn loose in its socket."

"But I don't understand."

"It will all be made clear at the inquest, no doubt. A most unfortunate thing for you, Mr. Meyer."

"Most unfortunate," declared the proprietor of the restaurant, extending his thick hands pathetically. "Most ruinous to my business."

"We will go upstairs now," said Harley. "You will kindly lead the way, Mr. Meyer, and the whole thing will be quite clear to you, Knox."

As the proprietor walked out of the office and upstairs to the second floor Harley whispered in my ear:

"Where did she go?"

"No.... Hamilton Place," I replied in an undertone.

116

"Good God!" muttered my friend, and clutched my arm so tightly that I winced. "Good God! The master touch, Knox! This crime was the work of a genius--of a genius with slightly, very slightly, oblique eyes."

Opening a door on the second landing, Mr. Meyer admitted us to a small supper-room. Its furniture consisted of a round dining table, several chairs, a couch, and very little else. I observed, however, that the furniture, carpet, and a few other appointments were of a character much more elegant than those of the public room below. A window which overlooked the street was open, so that the plush curtains which had been drawn aside moved slightly to and fro in the draught.

"The window of the tragedy, Knox," explained Harley.

He crossed the room.

"If you will stand here beside me you will see the gap in the railing caused by the breaking away of the fragment which now lies on Mr. Meyer's desk. Some few yards to the left in the street below is where the assault took place, of which we have heard, and the unfortunate Mr. De Lana, who was dining here alone--an eccentric custom of his-- naturally ran to the window upon hearing the disturbance and leaned out, supporting his weight upon the railing. The rail collapsed, and--we know the rest."

"It will ruin me," groaned Meyer; "it will give bad repute to my establishment."

"I fear it will," agreed Harley sympathetically, "unless we can manage to clear up one or two little difficulties which I have observed. For instance"--he tapped the proprietor on the shoulder confidentially --"have you any idea, any hazy idea, of the identity of the woman who was dining here with Mr. De Lana on Wednesday night?"

The effect of this simple inquiry upon the proprietor was phenomenal. His fat yellow face assumed a sort of leaden hue, and his already prominent eyes protruded abnormally. He licked his lips.

"I tell you--already I tell you," he muttered, "that Mr. De Lana he engage this room every Wednesday and sometimes also Friday, and dine here by himself."

"And I tell you," said Harley sweetly, "that you are an inspired liar. You smuggled her out by the side entrance after the accident."

"The side entrance?" muttered Meyer. "The side entrance?"

"Exactly; the side entrance. There is something else which I must ask you to tell me. Who had engaged this room on Tuesday night, the night before the accident?"

The proprietor's expression remained uncomprehending, and:

"A gentleman," he said. "I never see him before."

"Another solitary diner?" suggested Harley.

"Yes, he is alone all the evening waiting for a friend who does not arrive."

"Ah," mused Harley--"alone all the evening, was he? And his friend disappointed him. May I suggest that he was a dark man? Gray at the temples, having a dark beard and moustache, and a very tanned face? His eyes slanted slightly upward?"

117

"Yes! yes!" cried Meyer, and his astonishment was patently unfeigned. "It is a friend of yours?"

"A friend of mine, yes," said Harley absently, but his expression was very grim. "What time did he finally leave?"

"He waited until after eleven o'clock. The dinner is spoilt. He pays, but does not complain."

"No," said Harley musingly, "he had nothing to complain about. One more question, my friend. When the lady escaped hurriedly on Wednesday night, what was it that she left behind and what price are you trying to extort from her for returning it?"

At that the man collapsed entirely.

"Ah, Gott!" he cried, and raised his hand to his clammy forehead. "You will ruin me. I am a ruined man. I don't try to extort anything. I run an honest business..."

"And one of the most profitable in the world," added Harley, "since the days of Thais to our own. Even at Bond Street rentals I assume that a house of assignation is a golden enterprise."

"Ah!" groaned Meyer, "I am ruined, so what does it matter? I tell you everything. I know Mr. De Lana who engages my room regularly, but I don't know who the lady is who meets him here. No! I swear it! But always it is the same lady. When he falls I am downstairs in my office, and I hear him cry out. The lady comes running from the room and begs of me to get her away without being seen and to keep all mention of her out of the matter."

"What did she pay you?" asked Harley.

"Pay me?" muttered Meyer, pulled up thus shortly in the midst of his statement.

"Pay you. Exactly. Don't argue; answer."

The man delivered himself of a guttural, choking sound, and finally:

"She promised one hundred pounds," he confessed hoarsely.

"But you surely did not accept a mere promise? Out with it. What did she give you?"

"A ring," came the confession at last.

"A ring. I see. I will take it with me if you don't mind. And now, finally, what was it that she left behind?"

"Ah, Gott!" moaned the man, dropping into a chair and resting his arms upon the table. "It is all a great panic, you see. I hurry her out by the back stair from this landing and she forgets her bag."

"Her bag? Good."

"Then I clear away the remains of dinner so I can say Mr. De Lana is dining alone. It is as much my interest as the lady's."

"Of course! I quite understand. I will trouble you no more, Mr. Meyer, except to step into your office and to relieve you of that incriminating evidence, the lady's bag and her ring."

CHAPTER 4

THE SLANTING EYES

Do you understand, Knox?" said Harley as the cab bore us toward Hamilton Place. "Do you grasp the details of this cunning scheme?"

"On the contrary," I replied, "I am hopelessly at sea."

Nevertheless, I had forgotten that I was hungry in the excitement which now claimed me. For although the thread upon which these seemingly disconnected things hung was invisible to me, I recognized that Bampton, the city clerk, the bearded stranger who had made so singular a proposition to him, the white-hatted major, the dead stockbroker, and the mysterious woman whose presence in the case the clear sight of Harley had promptly detected, all were linked together by some subtle chain. I was convinced, too, that my friend held at least one end of that chain in his grip.

"In order to prepare your mind for the interview which I hope to obtain this evening," continued Harley, "let me enlighten you upon one or two points which may seem obscure. In the first place you recognize that anyone leaning out of the window on the second floor would almost automatically rest his weight upon the iron bar which was placed there for that very purpose, since the ledge is unusually low?"

"Quite," I replied, "and it also follows that if the bar gave way anyone thus leaning on it would be pitched into the street."

"Your reasoning is correct."

"But, my dear fellow," said I, "how could such an accident have been foreseen?"

"You speak of an accident. This was no accident! One end of the bar had been filed completely through, although the file marks had been carefully concealed with rust and dirt; and the other end had been wrenched out from its socket and then replaced in such a way that anyone leaning upon the bar could not fail to be precipitated into the street!"

"Good heavens! Then you mean..."

"I mean, Knox, that the man who occupied the supper room on the night before the tragedy--the dark man, tanned and bearded, with slightly oblique eyes---spent his time in filing through that bar--in short, in preparing a death trap!"

I was almost dumbfounded.

"But, Harley," I said, "assuming that he knew his victim would be the next occupant of the room, how could he know...?"

I stopped. Suddenly, as if a curtain had been raised, the details of what I now perceived to be a fiendishly cunning murder were revealed to me.

"According to his own account, Knox," resumed Harley, "Major Ragstaff regularly passed along that street with military punctuality at the same hour every night. You may take it for granted that the murderer was well aware of this. As a matter of fact, I happen to know that he was. We must also take it for granted that the murderer knew of these little dinners for two which took place in the private room above the Cafe Dame every Wednesday--and sometimes on Friday. Around the figure of the methodical major--with his conspicuous white hat as a sort of focus--was built up one of the most ingenious schemes of murder with which I have ever come in contact. The victim literally killed himself."

"But, Harley, the victim might have ignored the disturbance."

"That is where I first detected the touch of genius, Knox. He recognized the voice of one of the combatants--or his companion did. Here we are."

The cab drew up before the house in Hamilton Place. We alighted, and Harley pressed the bell. The same footman whom I had seen admit the woman opened the door.

"Is Lady Ireton at home?" asked Harley.

As he uttered the name I literally held my breath. We had come to the house of Major Ragstaff's daughter, the Marchioness of Ireton, one of society's most celebrated and beautiful hostesses!--the wife of a peer famed alike as sportsman, soldier, and scholar.

"I believe she is dining at home, sir," said the man. "Shall I inquire?"

"Be good enough to do so," replied Harley, and gave him a card. "Inform her that I wish to return to her a handbag which she lost a few days ago."

The man ushered us into an anteroom opening off the lofty and rather gloomy hall, and as the door closed:

"Harley," I said in a stage whisper, "am I to believe..."

"Can you doubt it?" returned Harley with a grim smile.

A few moments later we were shown into a charmingly intimate little boudoir in which Lady Ireton was waiting to receive us. She was a strikingly handsome brunette, but to-night her face, which normally, I think, possessed rich coloring, was almost pallid, and there was a hunted look in her dark eyes which made me wish to be anywhere rather than where I found myself. Without preamble she rose and addressed Harley:

"I fail to understand your message, sir," she said, and I admired the imperious courage with which she faced him. "You say you have recovered a handbag which I had lost?"

Harley bowed, and from the pocket of his greatcoat took out a silken-tasseled bag.

"The one which you left in the Cafe Dame, Lady Ireton," he replied. "Here also I have"-- from another pocket he drew out a diamond ring--"something which was extorted from you by the fellow Meyer."

Without touching her recovered property, Lady Ireton sank slowly down into the chair from which she had arisen, her gaze fixed as if hypnotically upon the speaker.

"My friend, Mr. Knox, is aware of all the circumstances," continued the latter, "but he is as anxious as I am to terminate this painful interview. I surmise that what occurred on Wednesday night was this--(correct me if I am wrong): While dining with Mr. De Lana you heard sounds of altercation in the street below. May I suggest that you recognized one of the voices?"

Lady Ireton, still staring straight before her at Harley, inclined her head in assent.

"I heard my father's voice," she said hoarsely.

"Quite so," he continued. "I am aware that Major Ragstaff is your father." He turned to me: "Do you recognize the touch of genius at last?" Then, again addressing Lady Ireton: "You naturally suggested to your companion that he should look out of the window in order to learn what was taking place. The next thing you knew was that he had fallen into the street below?"

Lady Ireton shuddered and raised her hands to her face.

"It is retribution," she whispered. "I have brought this ruin upon myself. But he does not deserve..."

Her voice faded into silence, and:

"You refer to your husband, Lord Ireton?" said Harley.

Lady Ireton nodded, and again recovering power of speech:

"It was to have been our last meeting," she said, looking up at Harley.

She shuddered, and her eyes blazed into sudden fierceness. Then, clenching her hands, she looked aside.

"Oh, God, the shame of this hour!" she whispered.

And I would have given much to have been spared the spectacle of this proud, erring woman's humiliation. But Paul Harley was scientifically remorseless. I could detect no pity in his glance.

"I would give my life willingly to spare my husband the knowledge of what has been," said Lady Ireton in a low, monotonous voice. "Three times I sent my maid to Meyer to recover my bag, but he demanded a price which even I could not pay. Now it is all discovered, and Harry will know."

"That, I fear, is unavoidable, Lady Ireton," declared Harley. "May I ask where Lord Ireton is at present?"

"He is in Africa after big game."

"H'm," said Harley, "in Africa, and after big game? I can offer you one consolation, Lady Ireton. In his own interests Meyer will stick to his first assertion that Mr. De Lana was dining alone."

A strange, horribly pathetic look came into the woman's haunted eyes.

"You--you--are not acting for...?" she began.

"I am acting for no one," replied Harley tersely. "Upon my friend's discretion you may rely as upon my own."

"Then why should he ever know?" she whispered.

"Why, indeed," murmured Harley, "since he is in Africa?"

As we descended the stair to the hall my friend paused and pointed to a life-sized oil painting by London's most fashionable portrait painter. It was that of a man in the uniform of a Guards officer, a dark man, slightly gray at the temples, his face very tanned as if by exposure to the sun.

"Having had no occasion for disguise when the portrait was painted," said Harley, "Lord Ireton appears here without the beard; and as he is not represented smiling one cannot see the gold tooth. But the painter, if anything, has accentuated the slanting eyes. You see, the fourth marquis--the present Lord Ireton's father--married one of the world-famous Yen Sun girls, daughters of the mandarin of that name by an Irish wife. Hence, the eyes. And hence..."

"But, Harley--it was murder!"

"Not within the meaning of the law, Knox. It was a recrudescence of Chinese humor! Lord Ireton is officially in Africa (and he went actually after 'big game'). The counsel is not born who could secure a conviction. We are somewhat late, but shall therefore have less difficulty in finding a table at Prince's."

TCHERIAPIN

CHAPTER I

THE ROSE

"Examine it closely," said the man in the unusual caped overcoat. "It will repay examination."

I held the little object in the palm of my hand, bending forward over the marble-topped table and looking down at it with deep curiosity. The babel of tongues so characteristic of Malay Jack's, and that mingled odor of stale spirits, greasy humanity, tobacco, cheap perfume, and opium, which distinguish the establishment faded from my ken. A sense of loneliness came to me.

Perhaps I should say that it became complete. I had grown conscious of its approach at the very moment that the cadaverous white-haired man had addressed me. There was a quality in his steadfast gaze and in his oddly pitched deep voice which from the first had wrapped me about--as though he were cloaking me in his queer personality and withdrawing me from the common plane.

Having stared for some moments at the object in my palm, I touched it gingerly; whereupon my acquaintance laughed--a short bass laugh.

"It looks fragile," he said. "But have no fear. It is nearly as hard as a diamond."

Thus encouraged, I took the thing up between finger and thumb, and held it before my eyes. For long enough I looked at it, and looking, my wonder grew. I thought that here was the most wonderful example of the lapidary's art which I had ever met with, east or west.

It was a tiny pink rose, no larger than the nail of my little finger. Stalk and leaves were there, and golden pollen lay in its delicate heart. Each fairy-petal blushed with June fire; the frail leaves were exquisitely green. Withal it was as hard and unbendable as a thing of steel.

"Allow me," said the masterful voice.

A powerful lens was passed by my acquaintance. I regarded the rose through the glass, and thereupon I knew, beyond doubt, that there was something phenomenal about the gem--if gem it were. I could plainly trace the veins and texture of every petal.

I suppose I looked somewhat startled. Although, baldly stated, the fact may not seem calculated to affright, in reality there was something so weird about this unnatural bloom that I dropped it on the table. As I did so I uttered an exclamation; for in spite of the stranger's assurances on the point, I had by no means overcome my idea of the thing's fragility.

"Don't be alarmed," he said, meeting my startled gaze. "It would need a steam-hammer to

123

do any serious damage."

He replaced the jewel in his pocket, and when I returned the lens to him he acknowledged it with a grave inclination of the head. As I looked into his sunken eyes, in which I thought lay a sort of sardonic merriment, the fantastic idea flashed through my mind that I had fallen into the clutches of an expert hypnotist who was amusing himself at my expense, that the miniature rose was a mere hallucination produced by the same means as the notorious Indian rope trick.

Then, looking around me at the cosmopolitan groups surrounding the many tables, and catching snatches of conversations dealing with subjects so diverse as the quality of whisky in Singapore, the frail beauty of Chinese maidens, and the ways of "bloody greasers," common sense reasserted itself.

I looked into the gray face of my acquaintance.

"I cannot believe," I said slowly, "that human ingenuity could so closely duplicate the handiwork of nature. Surely the gem is unique?--possibly one of those magical talismans of which we read in Eastern stories?"

My companion smiled.

"It is not a gem," he replied, "and while in a sense it is a product of human ingenuity, it is also the handiwork of nature."

I was badly puzzled, and doubtless revealed the fact, for the stranger laughed in his short fashion, and:

"I am not trying to mystify you," he assured me. "But the truth is so hard to believe sometimes that in the present case I hesitate to divulge it. Did you ever meet Tcheriapin?"

This abrupt change of topic somewhat startled me, but nevertheless:

"I once heard him play," I replied. "Why do you ask the question?"

"For this reason: Tcheriapin possessed the only other example of this art which so far as I am aware ever left the laboratory of the inventor. He occasionally wore it in his buttonhole."

"It is then a manufactured product of some sort?"

"As I have said, in a sense it is; but"--he drew the tiny exquisite ornament from his pocket again and held it up before me--"it is a natural bloom."

"What!"

"It is a natural bloom," replied my acquaintance, fixing his penetrating gaze upon me. "By a perfectly simple process invented by the cleverest chemist of his age it had been reduced to this gem-like state while retaining unimpaired every one of its natural beauties, every shade of its natural color. You are incredulous?"

"On the contrary," I replied, "having examined it through a magnifying glass I had already assured myself that no human hand had fashioned it. You arouse my curiosity intensely. Such a process, with its endless possibilities, should be worth a fortune to the inventor."

124

The stranger nodded grimly and again concealed the rose in his pocket.

"You are right," he said; "and the secret died with the man who discovered it--in the great explosion at the Vortex Works in 1917. You recall it? The T.N.T. factory? It shook all London, and fragments were cast into three counties."

"I recall it perfectly well."

"You remember also the death of Dr. Kreener, the chief chemist? He died in an endeavor to save some of the workpeople."

"I remember."

"He was the inventor of the process, but it was never put upon the market. He was a singular man, sir; as was once said of him--'A Don Juan of science.' Dame Nature gave him her heart unwooed. He trifled with science as some men trifle with love, tossing aside with a smile discoveries which would have made another famous. This"--tapping his breast pocket--"was one of them."

"You astound me. Do I understand you to mean that Dr. Kreener had invented a process for reducing any form of plant life to this condition?"

"Almost any form," was the guarded reply. "And some forms of animal life."

"What!"

"If you like"--the stranger leaned forward and grasped my arm--"I will tell you the story of Dr. Kreener's last experiment."

I was now intensely interested. I had not forgotten the heroic death of the man concerning whose work this chance acquaintance of mine seemed to know so much. And in the cadaverous face of the stranger as he sat there regarding me fixedly there was a promise and an allurement. I stood on the verge of strange things; so that, looking into the deep-set eyes, once again I felt the cloak being drawn about me, and I resigned myself willingly to the illusion.

From the moment when he began to speak again until that when I rose and followed him from Malay Jack's, as I shall presently relate, I became oblivious of my surroundings. I lived and moved through those last fevered hours in the lives of Dr. Kreener, Tcheriapin, the violinist, and that other tragic figure around whom the story centered. I append:

THE STRANGER'S STORY

I asked you (said the man in the caped coat) if you had ever seen Tcheriapin, and you replied that you had once heard him play. Having once heard him play you will not have forgotten him. At that time, although war still raged, all musical London was asking where he had come from and to what nation he belonged. Then when he disappeared it was variously reported, you will recall, that he had been shot as a spy and that he had escaped from England and was serving with the Austrian army. As to his parentage I can enlighten you in a measure. He was a Eurasian. His father was an aristocratic Chinaman, and his mother a Polish ballet-dancer--that was his parentage; but I would scarcely hesitate to affirm that he came from Hell; and I shall presently show you that he has certainly returned there.

125

You remember the strange stories current about him. The cunning ones said that he had a clever press agent. This was true enough. One of the most prominent agents in London discovered him playing in a Paris cabaret. Two months later he was playing at the Queen's Hall, and musical London lay at his feet.

He had something of the personality of Paganini, as you remember, except that he was a smaller man; long, gaunt, yellowish hands and the face of a haggard Mephistopheles. The critics quarreled about him, as critics only quarrel about real genius, and while one school proclaimed that Tcheriapin had discovered an entirely new technique, a revolutionary system of violin playing, another school was equally positive in declaring that he could not play at all, that he was a mountebank, a trickster, whose proper place was in a variety theatre.

There were stories, too, that were never published--not only about Tcheriapin, but concerning the Strad, upon which he played. If all this atmosphere of mystery which surrounded the man had truly been the work of a press agent, then the agent must have been as great a genius as his client. But I can assure you that the stories concerning Tcheriapin, true and absurd alike, were not inspired for business purposes; they grew up around him like fungi.

I can see him now, a lean, almost emaciated figure with slow, sinuous movements and a trick of glancing sideways with those dark, unfathomable, slightly oblique eyes. He could take up his bow in such a way as to create an atmosphere of electrical suspense.

He was loathsome, yet fascinating. One's mental attitude toward him was one of defense, of being tensely on guard. Then he would play.

You have heard him play, and it is therefore unnecessary for me to attempt to describe the effect of that music. The only composition which ever bore his name--I refer to "The Black Mass"--affected me on every occasion when I heard it, as no other composition has ever done.

Perhaps it was Tcheriapin's playing rather than the music itself which reached down into hitherto un-plumbed depths within me and awakened dark things which, unsuspected, lay there sleeping. I never heard "The Black Mass" played by anyone else; indeed, I am not aware that it was ever published. But had it been we should rarely hear it. Like Locke's music to "Macbeth" it bears an unpleasant reputation; to include it in any concert programme would be to court disaster. An idle superstition, perhaps, but there is much naivete in the artistic temperament.

Men detested Tcheriapin, yet when he chose he could win over his bitterest enemies. Women followed him as children followed the Pied Piper; he courted none, but was courted by all. He would glance aside with those black, slanting eyes, shrug in his insolent fashion, and turn away. And they would follow. God knows how many of them followed-- whether through the dens of Limehouse or the more fashionable salons of vice in the West End--they followed--perhaps down to Hell. So much for Tcheriapin.

At the time when the episode occurred to which I have referred, Dr. Kreener occupied a house in Regent's Park, to which, when his duties at the munition works allowed, he would sometimes retire at week-ends. He was a man of complex personality. I think no one ever knew him thoroughly; indeed, I doubt if he knew himself.

126

He was hail-fellow-well-met with the painters, sculptors, poets, and social reformers who have made of Soho a new Mecca. No movement in art was so modern that Dr. Kreener was not conversant with it; no development in Bolshevism so violent or so secret that Dr. Kreener could not speak of it complacently and with inside knowledge.

These were his Bohemian friends, these dreamers and schemers. Of this side of his life his scientific colleagues knew little or nothing, but in his hours of leisure at Regent's Park it was with these dreamers that he loved to surround himself rather than with his brethren of the laboratory. I think if Dr. Kreener had not been a great chemist he would have been a great painter, or perhaps a politician, or even a poet. Triumph was his birthright, and the fruits for which lesser men reached out in vain fell ripe into his hands.

The favorite meeting-place for these oddly assorted boon companions was the doctor's laboratory, which was divided from the house by a moderately large garden. Here on a Sunday evening one might meet the very "latest" composer, the sculptor bringing a new "message," or the man destined to supplant with the ballet the time-worn operatic tradition.

But while some of these would come and go, so that one could never count with certainty upon meeting them, there was one who never failed to be present when such an informal reception was held. Of him I must speak at greater length, for a reason which will shortly appear.

Andrews was the name by which he was known to the circles in which he moved. No one, from Sir John Tennier, the fashionable portrait painter, to Kruski, of the Russian ballet, disputed Andrews's right to be counted one of the elect. Yet it was known, nor did he trouble to hide the fact, that Andrews was employed at a large printing works in South London, designing advertisements. He was a great, red-bearded, unkempt Scotsman, and only once can I remember to have seen him strictly sober; but to hear him talk about painters and painting in his thick Caledonian accent was to look into the soul of an artist.

He was as sour as an unripe grape-fruit, cynical, embittered, a man savagely disappointed with life and the world; and tragedy was written all over him. If anyone knew the secret of his wasted life it was Dr. Kreener, and Dr. Kreener was a reliquary of so many secrets that this one was safe as if the grave had swallowed it.

One Sunday Tcheriapin joined the party. That he would gravitate there sooner or later was inevitable, for the laboratory in the garden was a Kaaba to which all such spirits made at least one pilgrimage. He had just set musical London on fire with his barbaric playing, and already those stories to which I have referred were creeping into circulation.

Although Dr. Kreener never expected anything of his guests beyond an interchange of ideas, it was a fact that the laboratory contained an almost unique collection of pencil and charcoal studies by famous artists, done upon the spot; of statuettes in wax, putty, soap and other extemporized materials, by the newest sculptors. While often enough from the drawing room which opened upon the other end of the garden had issued the strains of masterly piano-playing, and it was no uncommon thing for little groups to gather in the neighboring road to listen, gratis, to the voice of some great vocalist.

From the first moment of their meeting an intense antagonism sprang up between Tcheriapin and Andrews. Neither troubled very much to veil it. In Tcheriapin it found expression in covert sneers and sidelong glances, while the big, lion-maned Scotsman snorted open contempt of the Eurasian violinist. However, what I was about to say was that Tcheriapin on the occasion of his first visit brought his violin.

It was there, amid these incongruous surroundings, that I first had my spirit tortured by the strains of "The Black Mass."

There were five of us present, including Tcheriapin, and not one of the four listeners was unaffected by the music. But the influence which it exercised upon Andrews was so extraordinary as almost to reach the phenomenal. He literally writhed in his chair, and finally interrupted the performance by staggering rather than walking out of the laboratory.

I remember that he upset a jar of acid in his stumbling exit. It flowed across the floor almost to the feet of Tcheriapin, and the way in which the little black-haired man skipped, squealing, out of the path of the corroding fluid was curiously like that of a startled rabbit. Order was restored in due course, but we could not induce Tcheriapin to play again, nor did Andrews return until the violinist had taken his departure. We found him in the dining room, a nearly empty whisky-bottle beside him.

"I had to gang awa'," he explained thickly; "he was temptin' me to murder him. I should ha' had to do it if I had stayed. Damn his hell-music."

Tcheriapin revisited Dr. Kreener on many occasions afterward, although for a long time he did not bring his violin again. The doctor had prevailed upon Andrews to tolerate the Eurasian's company, and I could not help noticing how Tcheriapin skilfully and deliberately goaded the Scotsman, seeming to take a fiendish delight in disagreeing with his pet theories and in discussing any topic which he had found to be distasteful to Andrews.

Chief among these was that sort of irreverent criticism of women in which male parties so often indulge. Bitter cynic though he was, women were sacred to Andrews. To speak disrespectfully of a woman in his presence was like uttering blasphemy in the study of a cardinal. Tcheriapin very quickly detected the Scotsman's weakness, and one night he launched out into a series of amorous adventures which set Andrews writhing as he had writhed under the torture of "The Black Mass."

On this occasion the party was only a small one, comprising myself, Dr. Kreener, Andrews and Tcheriapin. I could feel the storm brewing, but was powerless to check it. How presently it was to break in tragic violence I could not foresee. Fate had not meant that I should foresee it.

Allowing for the free play of an extravagant artistic mind, Tcheriapin's career on his own showing had been that of a callous blackguard. I began by being disgusted and ended by being fascinated, not by the man's scandalous adventures, but by the scarcely human psychology of the narrator.

From Warsaw to Budapesth, Shanghai to Paris, and Cairo to London he passed, leaving ruin behind him with a smile--airily flicking cigarette ash upon the floor to indicate the termination of each "episode."

Andrews watched him in a lowering way which I did not like at all. He had ceased to snort his scorn; indeed, for ten minutes or so he had uttered no word or sound; but there was something in the pose of his ungainly body which strangely suggested that of a great dog preparing to spring. Presently the violinist recalled what he termed a "charming idyll of Normandy."

"There is one poor fool in the world," he said, shrugging his slight shoulders, "who never knew how badly he should hate me. Ha! ha! of him I shall tell you. Do you remember, my friends, some few years ago, a picture that was published in Paris and London? Everybody bought it; everybody said: 'He is a made man, this fellow who can paint so fine.'"

"To what picture do you refer?" asked Dr. Kreener.

"It was called 'A Dream at Dawn.'"

As he spoke the words I saw Andrews start forward, and Dr. Kreener exchanged a swift glance with him. But the Scotsman, unseen by the vainglorious half-caste, shook his head fiercely.

The picture to which Tcheriapin referred will, of course, be perfectly familiar to you. It had phenomenal popularity some eight years ago. Nothing was known of the painter-- whose name was Colquhoun--and nothing has been seen of his work since. The original painting was never sold, and after a time this promising new artist was, of course, forgotten.

Presently Tcheriapin continued:

"It is the figure of a slender girl--ah! angels of grace!--what a girl!" He kissed his hand rapturously. "She is posed bending gracefully forward, and looking down at her own lovely reflection in the water. It is a seashore, you remember, and the little ripples play about her ankles. The first blush of the dawn robes her white body in a transparent mantle of light. Ah! God's mercy! it was as she stood so, in a little cove of Normandy, that I saw her!"

He paused, rolling his dark eyes; and I could hear Andrews's heavy breathing; then:

"It was the 'new art'--the posing of the model not in a lighted studio, but in the scene to be depicted.

"And the fellow who painted her!--the man with the barbarous name! Bah! he was big--as big as our Mr. Andrews--and ugly--pooh! uglier than he! A moon-face, with cropped skull like a prize- fighter and no soul. But, yes, he could paint. 'A Dream at Dawn' was genius-- yes, some soul he must have had.

"He could paint, dear friends, but he could not love. Him I counted as--puff!"

He blew imaginary down into space.

"Her I sought out, and presently found. She told me, in those sweet stolen rambles along the shore, when the moonlight made her look like a Madonna, that she was his inspiration--his art--his life. And she wept; she wept, and I kissed her tears away.

"To please her I waited until 'A Dream at Dawn' was finished. With the finish of the picture, finished also his dream of dawn-- the moon-faced one's."

129

Tcheriapin laughed, and lighted a fresh cigarette.

"Can you believe that a man could be so stupid? He never knew of my existence, this big, red booby. He never knew that I existed until--until his 'dream' had fled--with me! In a week we were in Paris, that dream-girl and I--in a month we had quarreled. I always end these matters with a quarrel; it makes the complete finish. She struck me in the face--and I laughed. She turned and went away. We were tired of one another.

"Ah!" Again he airily kissed his hand. "There were others after I had gone. I heard for a time. But her memory is like a rose, fresh and fair and sweet. I am glad I can remember her so, and not as she afterward became. That is the art of love. She killed herself with absinthe, my friends. She died in Marseilles in the first year of the great war."

Thus far Tcheriapin had proceeded, and was in the act of airily flicking ash upon the floor, when, uttering a sound which I can only describe as a roar, Andrews hurled himself upon the smiling violinist.

His great red hands clutching Tcheriapin's throat, the insane Scotsman, for insane he was at that moment, forced the other back upon the settee from which he had half arisen. In vain I sought to drag him away from the writhing body, but I doubt that any man could have relaxed that deadly grip. Tcheriapin's eyes protruded hideously and his tongue lolled forth from his mouth. One could hear the breath whistling through his nostrils as Andrews silently, deliberately, squeezed the life out of him.

It all occupied only a few minutes, and then Andrews, slowly opening his rigidly crooked fingers, stood panting and looking down at the distorted face of the dead man.

For once in his life the Scotsman was sober, and turning to Dr. Kreener:

"I have waited seven long years for this," he said, "and I'll hang wi' contentment."

I can never forget the ensuing moments, in which, amid a horrible silence broken only by the ticking of a clock and the heavy breathing of Colquhoun (so long known to us as Andrews) we stood watching the contorted body on the settee.

And as we watched, slowly the rigid limbs began to relax, and Tcheriapin slid gently on to the floor, collapsing there with a soft thud, where he squatted like some hideous Buddha, resting back against the cushions, one spectral yellow hand upraised, the fingers still clutching a big gold tassel.

Andrews (for so I always think of him) was seized with a violent fit of trembling, and he dropped into the chair, muttering to himself and looking down wild-eyed at his twitching fingers. Then he began to laugh, high-pitched laughter, in little short peals.

"Here!" cried the doctor sharply. "Drop that!"

Crossing to Andrews, he grasped him by the shoulders and shook him roughly.

The laughter ceased, and:

"Send for the police," said Andrews in a queer, shaky voice. "Dinna fear but I'm ready. I'm only sorry it happened here."

"You ought to be glad," said Dr. Kreener.

There was a covert meaning in the words--a fact which penetrated even to the dulled intelligence of the Scotsman, for he glanced up haggardly at his friend.

"You ought to be glad," repeated Dr. Kreener.

Turning, he walked to the laboratory door and locked it. He next lowered all the blinds.

"I pray that we have not been observed," he said, "but we must chance it."

He mixed a drink for Andrews and himself. His quiet, decisive manner had had its effect, and Andrews was now more composed. Indeed, he seemed to be in a half-dazed condition; but he persistently kept his back turned to the crouching figure propped up against the settee.

"If you think you can follow me," said Dr. Kreener abruptly, "I will show you the result of a recent experiment."

Unlocking a cupboard, he took out a tiny figure some two inches long by one inch high, mounted upon a polished wooden pedestal. It was that of a guinea-pig. The flaky fur gleamed like the finest silk, and one felt that the coat of the minute creature would be as floss to the touch; whereas in reality it possessed the rigidity of steel. Literally one could have done it little damage with a hammer. Its weight was extraordinary.

"I am learning new things about this process every day," continued Dr. Kreener, placing the little figure upon a table. "For instance, while it seems to operate uniformly upon vegetable matter, there are curious modifications when one applies it to animal and mineral substances. I have now definitely decided that the result of this particular inquiry must never be published. You, Colquhoun, I believe, possess an example of the process, a tiger lily, I think? I must ask you to return it to me. Our late friend, Tcheriapin, wears a pink rose in his coat which I have treated in the same way. I am going to take the liberty of removing it."

He spoke in the hard, incisive manner which I had heard him use in the lecture theatre, and it was evident enough that his design was to prepare Andrews for something which he contemplated. Facing the Scotsman where he sat hunched up in the big armchair, dully watching the speaker:

"There is one experiment," said Dr. Kreener, speaking very deliberately, "which I have never before had a suitable opportunity of attempting. Of its result I am personally confident, but science always demands proof."

His voice rang now with a note of repressed excitement. He paused for a moment, and then:

"If you were to examine this little specimen very closely," he said, and rested his finger upon the tiny figure of the guinea- pig, "you would find that in one particular it is imperfect. Although a diamond drill would have to be employed to demonstrate the fact, the animal's organs, despite their having undergone a chemical change quite new to science, are intact, perfect down to the smallest detail. One part of the creature's structure alone defied my process. In short, dental enamel is impervious to it. This little animal, otherwise as complete as when it lived and breathed, has no teeth. I found it necessary to extract them before submitting the body to the reductionary process."

131

He paused.

"Shall I go on?" he asked.

Andrews, to whose mind, I think, no conception of the doctor's project had yet penetrated, shuddered, but slowly nodded his head.

Dr. Kreener glanced across the laboratory at the crouching figure of Tcheriapin, then, resting his hands upon Andrews's shoulders, he pushed him back in the chair and stared into his dull eyes.

"Brace yourself, Colquhoun," he said tersely.

Turning, he crossed to a small mahogany cabinet at the farther end of the room. Pulling out a glass tray he judicially selected a pair of dental forceps.

CHAPTER 2

"THE BLACK MASS"

Thus far the stranger's appalling story had progressed when that singular cloak in which hypnotically he had enwrapped me seemed to drop, and I found myself clutching the edge of the table and staring into the gray face of the speaker.

I became suddenly aware of the babel of voices about me, of the noisome smell of Malay Jack's, and of the presence of Jack in person, who was inquiring if there were any further orders. I was conscious of nausea.

"Excuse me," I said, rising unsteadily, "but I fear the oppressive atmosphere is affecting me."

"If you prefer to go out," said my acquaintance, in that deep voice which throughout the dreadful story had rendered me oblivious of my surroundings, "I should be much favored if you would accompany me to a spot not five hundred yards from here."

Seeing me hesitate:

"I have a particular reason for asking," he added.

"Very well," I replied, inclining my head, "if you wish it. But certainly I must seek the fresh air."

Going up the steps and out through the door above which the blue lantern burned, we came to the street, turned to the left, to the left again, and soon were threading that maze of narrow ways which complicates the map of Pennyfields.

I felt somewhat recovered. Here, in the narrow but familiar highways the spell of my singular acquaintance lost much of its potency, and already I found myself doubting the story of Dr. Kreener and Tcheriapin. Indeed, I began to laugh at myself, conceiving that I had fallen into the hands of some comedian who was making sport of me; although why such a person should visit Malay Jack's was not apparent.

132

I was about to give expression to these new and saner ideas when my companion paused before a door half hidden in a little alley which divided the back of a Chinese restaurant from the tawdry- looking establishment of a cigar merchant. He apparently held the key, for although I did not actually hear the turning of the lock I saw that he had opened the door.

"May I request you to follow me?" came his deep voice out of the darkness. "I will show you something which will repay your trouble."

Again the cloak touched me, but it was without entirely resigning myself to the compelling influence that I followed my mysterious acquaintance up an uncarpeted and nearly dark stair. On the landing above a gas lamp was burning, and opening a door immediately facing the stair the stranger conducted me into a barely furnished and untidy room.

The atmosphere smelled like that of a pot-house, the odors of stale spirits and of tobacco mingling unpleasantly. As my guide removed his hat and stood there, a square, gaunt figure in his queer, caped overcoat, I secured for the first time a view of his face in profile; and found it to be startlingly unfamiliar. Seen thus, my acquaintance was another man. I realized that there was something unnatural about the long, white hair, the gray face; that the sharp outline of brow, nose, and chin was that of a much younger man than I had supposed him to be.

All this came to me in a momentary flash of perception, for immediately my attention was riveted upon a figure hunched up on a dilapidated sofa on the opposite side of the room. It was that of. a big man, bearded and very heavily built, but whose face was scarred as by years of suffering, and whose eyes confirmed the story indicated by the smell of stale spirits with which the air of the room was laden. A nearly empty bottle stood on a table at his elbow, a glass beside it, and a pipe lay in a saucer full of ashes near the glass.

As we entered, the glazed eyes of the man opened widely and he clutched at the table with big red hands, leaning forward and staring horribly.

Save for this derelict figure and some few dirty utensils and scattered garments which indicated that the apartment was used both as sleeping and living room, there was so little of interest in the place that automatically my wandering gaze strayed from the figure on the sofa to a large oil painting, unframed, which rested upon the mantelpiece above the dirty grate, in which the fire had become extinguished.

I uttered a stifled exclamation. It was "A Dream at Dawn"-- evidently the original painting!

On the left of it, from a nail in the wall, hung a violin and bow, and on the right stood a sort of cylindrical glass case or closed jar, upon a wooden base.

From the moment that I perceived the contents of this glass case a sense of fantasy claimed me, and I ceased to know where reality ended and mirage began.

It contained a tiny and perfect figure of a man. He was arrayed in a beautifully fitting dress-suit such as a doll might have worn, and he was posed as if in the act of playing a violin, although no violin was present. At the elfin black hair and Mephistophelian face of this horrible, wonderful image, I stared fascinatedly.

I looked and looked at the dwarfed figure of. . . Tcheriapin!

All these impressions came to me in the space of a few hectic moments, when in upon my mental tumult intruded a husky whisper from the man on the sofa.

"Kreener!" he said. "Kreener!"

At the sound of that name, and because of the way in which it was pronounced, I felt my blood running cold. The speaker was staring straight at my companion.

I clutched at the open door. I felt that there was still some crowning horror to come. I wanted to escape from that reeking room, but my muscles refused to obey me, and there I stood while:

"Kreener!" repeated the husky voice, and I saw that the speaker was rising unsteadily to his feet.

"You have brought him again. Why have you brought him again? He will play. He will play me a step nearer to Hell."

"Brace yourself, Colquhoun," said the voice of my companion. "Brace yourself."

"Take him awa'!" came in a sudden frenzied shriek. "Take him awa'! He's there at your elbow, Kreener, mockin' me, and pointing to that damned violin."

"Here!" said the stranger, a high note of command in his voice. "Drop that! Sit down at once."

Even as the other obeyed him, the cloaked stranger, stepping to the mantelpiece, opened a small box which lay there beside the glass case. He turned to me; and I tried to shrink away from him. For I knew--I knew--yet I loathed to look upon--what was in the box. Muffled as though reaching me through fog, I heard the words:

"A perfect human body...in miniature. . . every organ intact by means of. . . process. . . rendered indestructible. Tcheriapin as he was in life may be seen by the curious ten thousand years hence. Incomplete. . . one respect. . . here in this box. . ."

The spell was broken by a horrifying shriek from the man whom my companion had addressed as Colquhoun, and whom I could only suppose to be the painter of the celebrated picture which rested upon the mantelshelf.

"Take him awa', Kreener! He is reaching for the violin!"

Animation returned to me, and I fell rather than ran down the darkened stair. How I opened the street door I know not, but even as I stepped out into the squalid alleys of Pennyfields the cloaked figure was beside me. A hand was laid upon my shoulder.

"Listen!" commanded a deep voice.

Clearly, with an eerie sweetness, an evil, hellish beauty indescribable, the wailing of a Stradivarius violin crept to my ears from the room above. Slowly--slowly the music began, and my soul rose up in revolt.

"Listen!" repeated the voice. "Listen! It is 'The Black Mass'!"

THE DANCE OF THE VEILS

CHAPTER I

THE HOUSE OF THE AGAPOULOS

Hassan came in and began very deliberately to light the four lamps. He muttered to himself and often smiled in the childish manner which characterizes some Egyptians. Hassan wore a red cap, and a white robe confined at the waist by a red sash. On his brown feet he wore loose slippers, also of red. He had good features and made a very picturesque figure moving slowly about his work.

As he lighted lamp after lamp and soft illumination crept about the big room, because of the heavy shadows created the place seemed to become mysteriously enlarged. That it was an Eastern apartment cunningly devised to appeal to the Western eye, one familiar with Arab households must have seen at once. It was a traditional Oriental interior, a stage setting rather than the nondescript and generally uninteresting environment of the modern Egyptian at home.

Brightly colored divans there were and many silken cushions of strange pattern and design. The hanging lamps were of perforated brass with little colored glass panels. In carved wooden cabinets stood beautiful porcelain jars, trays, and vessels of silver and copper ware. Rich carpets were spread about the floor, and the draperies were elegant and costly, while two deep windows projecting over the court represented the best period of Arab architecture. Their intricate carven woodwork had once adorned the palace of a Grand Wazir. Agapoulos had bought them in Cairo and had had them fitted to his house in Chinatown. A smaller brass lamp of very delicate workmanship was suspended in each of the recesses.

As Hassan, having lighted the four larger lanterns, was proceeding leisurely to light the first of the smaller ones, draperies before a door at the east end of the room were parted and Agapoulos came in. Agapoulos was a short but portly Greek whom the careless observer might easily have mistaken for a Jew. He had much of the appearance of a bank manager, having the manners of one used to making himself agreeable, but also possessing the money-eye and that comprehensive glance which belongs to the successful man of commerce.

Standing in the centre of the place he brushed his neat black moustache with a plump forefinger. A diamond ring which he wore glittered brilliantly in the colored rays of the lanterns. With his right hand, which rested in his trouser pocket, he rattled keys. His glance roved about the room appraisingly. Walking to a beautifully carved Arab cabinet he rearranged three pieces of Persian copperware which stood upon it. He moved several cushions, and taking up a leopard skin which lay upon the floor he draped it over an ebony chair which was inlaid intricately with ivory.

135

The drooping eyelids of M. Agapoulos drooped lower, as returning to the centre of the room he critically surveyed the effect of these master touches. At the moment he resembled a window- dresser, or, rather, one of those high-salaried artists who beautify the great establishments of Regent Street, the Rue de la Paix, and Ruination Avenue, New York.

Hassan lighted the sixth lamp, muttering smilingly all the time. He was about to depart when Agapoulos addressed him in Arabic.

"There will be a party down from the Savoy tonight, Hassan. No one else is to come unless I am told. That accursed red policeman, Kerry, has been about here of late. Be very careful."

Hassan saluted him gravely and retired through one of the draped openings. In his hand he held the taper with which he had lighted the lamps. In order that the draperies should not be singed he had to hold them widely apart. For it had not occurred to Hassan to extinguish the taper. The Egyptian mind is complex in its simplicity.

M. Agapoulos from a gold case extracted a cigarette, and lighting it, inhaled the smoke contentedly, looking about him. The window-dresser was lost again in the bank manager who has arranged a profitable overdraft. Somewhere a bell rang. Hassan, treading silently, reappeared, crossed the room, and opening a finely carved door walked along a corridor which it had concealed. He still carried the lighted taper.

Presently there entered a man whose well-cut serge suit revealed the figure of a soldier. He wore a soft gray felt hat and carried light gloves and a cane. His dark face, bronzed by recent exposure to the Egyptian sun, was handsome in a saturnine fashion, and a touch of gray at the temples tended to enhance his good looks. He carried himself in that kind of nonchalant manner which is not only insular but almost insolent.

M. Agapoulos bowed extravagantly. As he laid his plump hand upon his breast the diamond ring sparkled in a way most opulent and impressive.

"I greet you, Major Grantham," he said. "Behold"--he waved his hand glitteringly--"all is prepared."

"Oh, yes," murmured the other, glancing around without interest; "good. You are beginning to get straight in your new quarters."

Agapoulos extended the prosperous cigarette-case, and Major Grantham took and lighted a superior cigarette.

"How many in the party?" inquired the Greek smilingly.

"Three and myself."

A shadow of a frown appeared upon the face of Agapoulos.

"Only three," he muttered.

Major Grantham laughed.

"You should know me by this time, Agapoulos," he said. "The party is small but exclusive, you understand?"

He spoke wearily, as a tired man speaks of distasteful work which he must do. There was contempt in his voice; contempt of Agapoulos, and contempt of himself.

"Ah!" cried the Greek, brightening; "do I know any of them?"

"Probably. General Sir Francis Payne, Mr. Eddie, and Sir Horace Tipton."

"An Anglo-American party, eh?"

"Quite. Mr. Eddie is the proprietor of the well-known group of American hotels justly celebrated for their great height and poisonous cuisine; while Sir Horace Tipton alike as sportsman, globe-trotter, and soap manufacturer, is characteristically British. Of General Sir Francis Payne I need only say that his home services during the war did incalculable harm to our prestige throughout the Empire."

He spoke with all the bitterness of a man who has made a failure of life. Agapoulos was quite restored to good humor.

"Ah!" he exclaimed, brushing his moustache and rattling his keys; "sportsmen, eh?"

Major Grantham dropped into the carven chair upon which the Greek had draped the leopard skin. Momentarily the window-dresser leapt into life as Agapoulos beheld one of his cunning effects destroyed, but he forced a smile when Grantham, shrugging his shoulders, replied:

"If they are fools enough to play--the usual 5 per cent, on the bank's takings."

He paused, glancing at some ash upon the tip of his cigarette. Agapoulos swiftly produced an ashtray and received the ash on it in the manner of a churchwarden collecting half a crown from a pew-holder.

"I think," continued Grantham indifferently, "that it will be the dances. Two of them are over fifty."

"Ah!" said Agapoulos thoughtfully; "not, of course, the ordinary programme?"

Major Grantham looked up at him with lazy insolence.

"Why ask?" he inquired. "Does Lucullus crave for sausages? Do philosophers play marbles?"

He laughed again, noting the rather blank look of Agapoulos.

"You don't know what I'm talking about, do you?" he added. "I mean to say that these men have been everywhere and done everything. They have drunk wine sweet and sour and have swallowed the dregs. I am bringing them. It is enough."

"More than enough," declared the Greek with enthusiasm. He bowed, although Grantham was not looking at him. "In the little matter of fees I can rely upon your discretion, as always. Is it not said that a good dragoman is a desirable husband?"

Major Grantham resettled himself in his chair.

"M. Agapoulos," he said icily, "we have done shady business together for years, both in Port Said and in London, and have remained the best of friends; two blackguards linked by our common villainy. But if this pleasant commercial acquaintance is to continue let there be no misunderstanding between us, M. Agapoulos. I may know I'm a dragoman; but in future, old friend"--he turned lazy eyes upon the Greek--"for your guidance, don't remind me of the fact or I'll wring your neck."

The drooping eyelids of M. Agapoulos flickered significantly, but it was with a flourish more grand than usual that he bowed.

"Pardon, pardon," he murmured. "You speak harshly of yourself, but ah, you do not mean it. We understand each other, eh?"

"I understand you perfectly," drawled Grantham; "I was merely advising you to endeavor to understand me. My party will arrive at nine o'clock, Agapoulos, and I am going back to the Savoy shortly to dress. Meanwhile, if Hassan would bring me a whisky and soda I should be obliged."

"Of course, of course. He shall do so at once," cried Agapoulos. "I will tell him."

Palpably glad to escape, the fat Greek retired, leaving Major Grantham lolling there upon the leopard skin, his hat, cane and gloves upon the carpet beside him; and a few moments later Hassan the silent glided into the extravagant apartment bearing refreshments. Placing his tray upon a little coffee-table beside Major Grantham, he departed.

There was a faint smell of perfume in the room, a heavy voluptuous smell in which the odor of sandal-wood mingled with the pungency of myrrh. It was very silent, so that when Grantham mixed a drink the pleasant chink of glass upon glass rang out sharply.

CHAPTER 2

ZAHARA

Zahara had overheard the latter part of the conversation from her own apartment. Once she had even crept across to the carven screen in order that she might peep through into the big, softly lighted room. She had interrupted her toilet to do so, and having satisfied herself that Grantham was one of the speakers (although she had really known this already), she had returned and stared at herself critically in the mirror.

Zahara, whose father had been a Frenchman, possessed skin of a subtle cream color very far removed from the warm brown of her Egyptian mother, but yet not white. At night it appeared dazzling, for she enhanced its smooth, creamy pallor with a wonderful liquid solution which came from Paris. It was hard, Zahara had learned, to avoid a certain streaky appearance, but much practice had made her an adept.

This portion of her toilet she had already completed and studying her own reflection she wondered, as she had always wondered, what Agapoulos could see in Safiyeh. Safiyeh was as brown as a berry; quite pretty for an Egyptian girl, as Zahara admitted scornfully, but brown--brown. It was a great puzzle to Zahara. The mystery of life indeed had puzzled little Zahara very much from the moment when she had first begun to notice things with those big, surprising blue eyes of hers, right up to the present twenty- fourth year of her life. She had an uneasy feeling that Safiyeh, who was only sixteen, knew more of this mystery than she did. Once, shortly after the Egyptian girl had come to the house of

Agapoulos, Zahara had playfully placed her round white arm against that of the more dusky beauty, and:

"Look!" she had exclaimed. "I am cream and you are coffee."

"It is true," the other had admitted in her practical, serious way, "but some men do not like cream. All men like coffee."

Zahara rested her elbows upon the table and surveyed the reflection of her perfect shoulders with disapproval. She had been taught at her mother's knee that men did not understand women, and she, who had been born and reared in that quarter of Cairo where there is no day but one long night, had lived to learn the truth of the lesson. Yet she was not surprised that this was so; for Zahara did not understand herself. Her desires were so simple and so seemingly natural, yet it would appear that they were contrary to the established order of things.

She was proud to think that she was French, although someone had told her that the French, though brave, were mercenary. Zahara admired the French for being brave, and thought it very sensible that they should be mercenary. For there was nothing that Zahara wanted of the world that money could not obtain (or so she believed), and she knew no higher philosophy than the quest of happiness. Because others did not seem to share this philosophy she often wondered if she could be unusual. She had come to the conclusion that she was ignorant. If only Harry Grantham would talk to her she felt sure he could teach her so much.

There were so many things that puzzled her. She knew that at twenty-four she was young for a French girl, although as an Egyptian she would have been considered old. She had been taught that gold was the key to happiness and that man was the ogre from whom this key must be wheedled. A ready pupil, Zahara had early acquired the art of attracting, and now at twenty-four she was a past mistress of the Great Craft, and as her mirror told her, more beautiful than she had ever been.

Therefore, what did Agapoulos see in Safiyeh?

It was a problem which made Zahara's head ache. She could not understand why as her power of winning men increased her power to hold them diminished. Safiyeh was a mere inexperienced child-- yet Agapoulos had brought her to the house, and Zahara, wise in woman's lore, had recognized the familiar change of manner.

It was a great problem, the age-old problem which doubtless set the first silver thread among Phryne's red-gold locks and which now brought a little perplexed wrinkle between Zahara's delicately penciled brows.

It had not always been so. In those early days in Cairo there had been an American boy. Zahara had never forgotten. Her beauty had bewildered him. He had wanted to take her to New York; and oh! how she had wanted to go. But her mother, who was then alive, had held other views, and he had gone alone. Heavens! How old she felt. How many had come and gone since that Egyptian winter, but now, although admiration was fatally easy to win how few were so sincere as that fresh-faced boy from beyond the Atlantic.

Zahara, staring into the mirror, observed that there was not a wrinkle upon her face, not a flaw upon her perfect skin. Nor in this was she blinded by vanity. Nature, indeed, had cast her in a rare mould, and from her unusual hair, which was like dull gold, to her slender ankles and tiny feet, she was one of the most perfectly fashioned human beings who ever added to the beauty of the world.

Yet Agapoulos preferred Safiyeh. Zahara could hear him coming to her room even as she sat there, chin in hands, staring at her own bewitching reflection. Presently she would slip out and speak to Harry Grantham. Twice she had read in his eyes that sort of interest which she knew so well how to detect. She liked him very much, but because of a sense of loyalty to Agapoulos (a sentiment purely Egyptian which she longed to crush) Zahara had never so much as glanced at Grantham in the Right Way. She was glad, though, that he had not gone, and she hoped that Agapoulos would not detain her long.

As a matter of fact, the Greek's manner was even more cold than usual. He rested his hand upon her shoulder for a moment, and meeting her glance reflected in the mirror:

"There will be a lot of money here to-night," he said. "Make the best of your opportunities. Chinatown is foggy, yes--but it pays better than Port Said."

He ran fat fingers carelessly through her hair, the big diamond glittering effectively in the wavy gold, then turned and went out. Sitting listening intently, Zahara could hear him talking in a subdued voice to Safiyeh, and could detect the Egyptian's low-spoken replies.

*****

Grantham looked up with a start. A new and subtle perfume had added itself to that with which the air of the room was already laden. He found Zahara standing beside him.

His glance traveled upward from a pair of absurdly tiny brocaded shoes past slender white ankles to the embroidered edge of a wonderful mandarin robe decorated with the figures of peacocks; upward again to a little bejewelled hand which held the robe confined about the slender figure of Zahara, and upward to where, sideways upon a bare shoulder peeping impudently out from Chinese embroidery, rested the half-mocking and half-serious face of the girl.

"Hallo!" he said, smiling, "I didn't hear you come in."

"I walk very soft," explained Zahara, "because I am not supposed to be here."

She looked at him quizzically. "I don't see you for a long time," she added, and in the tone of her voice there was a caress. "I saw you more often in Port Said than here."

"No," replied Grantham, "I have been giving Agapoulos a rest. Besides, there has been nobody worth while at any of the hotels or clubs during the last fortnight."

"Somebody worth while coming to-night?" asked Zahara with professional interest.

At the very moment that she uttered the words she recognized her error, for she saw Grantham's expression change. Yet to her strange soul there was a challenge in his coldness and the joy of contest in the task of melting the ice of this English reserve.

"Lots of money," he said bitterly; "we shall all do well to- night."

Zahara did not reply for a moment. She wished to close this line of conversation which inadvertently she had opened up. So that, presently:

140

"You look very lonely and bored," she said softly.

As a matter of fact, it was she who was bored of the life she led in Limehouse--in chilly, misty Limehouse--and who had grown so very lonely since Safiyeh had come. In the dark gray eyes looking up at her she read recognition of her secret. Here was a man possessing that rare masculine attribute, intuition. Zahara knew a fear that was half delightful. Fear because she might fail in either of two ways and delight because the contest was equal.

"Yes," he replied slowly, "my looks tell the truth. How did you know?"

Zahara observed that his curiosity had not yet become actual interest. She toyed with the silken tassel on her robe, tying and untying it with quick nervous fingers and resting the while against the side of the carved chair.

"Perhaps because I am so lonely myself," she said. "I matter to no one. What I do, where I go, if I live or die. It is all----"

She spread her small hands eloquently and shrugged so that another white shoulder escaped from the Chinese wrapping. Thereupon Zahara demurely drew her robe about her with a naive air of modesty which nine out of ten beholding must have supposed to be affected.

In reality it was a perfectly natural, instinctive movement. To Zahara her own beauty was a commonplace to be displayed or concealed as circumstances might dictate. In a certain sense, which few could appreciate, this half-caste dancing girl and daughter of El Wasr was as innocent as a baby. It was one of the things which men did not understand. She thought that if Harry Grantham asked her to go away with him it would be nice to go. Suddenly she realized how deep was her loathing of this Limehouse and of the people she met there, who were all alike.

He sat looking at her for some time, and then: "Perhaps you are wrong," he said. "There may be some who could understand."

And because he had answered her thoughts rather than her words, the fear within Zahara grew greater than the joy of the contest.

Awhile longer she stayed, seeking for a chink in the armor. But she failed to kindle the light in his eyes which--unless she had deluded herself--she had seen there in the past; and because she failed and could detect no note of tenderness in his impersonal curiosity:

"You are lonely because you are so English, so cold," she exclaimed, drawing her robe about her and glancing sideways toward the door by which Agapoulos might be expected to enter. "You are bored, yes. Of course. You look on at life. It is not exciting, that game-- except for the players."

Never once had she looked at him in the Right Way; for to have done so and to have evoked only that amused yet compassionate smile would have meant hatred, and Zahara had been taught that such hatred was fatal because it was a confession of defeat.

"I shall see you again to-night, shall I not?" he said as she turned away.

"Oh, yes, I shall be--on show. I hope you will approve."

She tossed her head like a petulant child, turned, and with never another glance in his

141

direction, walked from the room. She was very graceful, he thought.

Yet it was not entirely of this strange half-caste, whose beauty was provoking, although he resolutely repelled her tentative advances, that Grantham was thinking. In that last gesture when she had scornfully tossed her head in turning aside, had lain a bitter memory. Grantham stood for a moment watching the swaying draperies. Then, dropping the end of his cigarette into a little brass ash-tray, he took up his hat, gloves, and cane from the floor, and walked toward the doorway through which he had entered.

A bell rang somewhere, and Grantham paused. A close observer might have been puzzled by his expression. Evidently changing his mind, he crossed the room, opened the door and went out, leaving the house of Agapoulos by a side entrance. Crossing the little courtyard below he hurried in the direction of the main street, seeming to doubt the shadows which dusk was painting in the narrow ways.

Many men who know Chinatown distrust its shadows, but the furtive fear of which Grantham had become aware was due not to anticipation but to memory--to a memory conjured up by that gesture of Zahara's.

There were few people in London or elsewhere who knew the history of this scallywag Englishman. That he had held the King's commission at some time was generally assumed to be the fact, but that his real name was not Grantham equally was taken for granted. His continuing, nevertheless, to style himself "Major" was sufficient evidence to those interested that Grantham lived by his wits; and from the fact that he lived well and dressed well one might have deduced that his wits were bright if his morals were turbid.

Now, the gesture of a woman piqued had called up the deathless past. Hurrying through nearly empty squalid streets, he found himself longing to pronounce a name, to hear it spoken that he might linger over its bitter sweetness. To this longing he presently succumbed, and:

"Inez," he whispered, and again more loudly, "Inez."

Such a wave of lonely wretchedness and remorse swept up about his heart that he was almost overwhelmed by it, yet he resigned himself to its ruthless cruelty with a sort of savage joy. The shadowed ways of Limehouse ceased to exist for him, and in spirit he stood once more in a queer, climbing, sunbathed street of Gibraltar looking out across that blue ribbon of the Straits to where the African coast lay hidden in the haze.

"I never knew," he said aloud. And one meeting this man who hurried along and muttered to himself must have supposed him to be mad. "I never knew. Oh, God! if I had only known."

But he was one of those to whom knowledge comes as a bitter aftermath. When his regiment had received orders to move from the Rock, and he had informed Inez of his departure, she had turned aside, just as Zahara had done; scornfully and in silence. Because of his disbelief in her he had guarded his heart against this beautiful Spanish girl who (as he realized too late) had brought him the only real happiness he had ever known. Often she had told him of her brother, Miguel, who would kill her--would kill them both--if he so much as suspected their meetings; of her affianced husband, absent in Tunis, whose jealousy knew no bounds.

142

He had pretended to believe, had even wanted to believe; but the witchery of the girl's presence removed, he had laughed--at himself and at Inez. She was playing the Great Game, skilfully, exquisitely. When he was gone--there would soon be someone else. Yet he had never told her that he doubted. He had promised many things--and had left her.

She died by her own hand on the night of his departure.

Now, as a wandering taxi came into view: "Inez!" he moaned--"I never knew."

That brother whom he had counted a myth had succeeded in getting on board the transport. Before Grantham's inner vision the whole dreadful scene now was reenacted: the struggle in the stateroom; he even seemed to hear the sound of the shot, to see the Spaniard, drenched with blood from a wound in his forehead, to hear his cry:

"I cannot see! I cannot see! Mother of Mercy! I have lost my sight!"

It had broken Grantham. The scandal was hushed up, but retirement was inevitable. He knew, too, that the light had gone out of the world for him as it had gone for Miguel da Mura.

It is sometimes thus that a scallywag is made.

CHAPTER 4

THE STAR OF EGYPT

As Grantham went out by the side door, Hassan, soft of foot, appeared. Crossing to the main door he opened it and walked down the narrow corridor beyond. Presently came the tap, tap, tap of a stick and a sound of muttered conversation in some place below.

Hassan reentered and went in through the curtained doorway to summon Agapoulos. Agapoulos was dressing and would not be disturbed. Hassan went back to those who waited, but ere long returned again chattering volubly to himself. Going behind the carven screen he rapped upon the door of Zahara's room, and she directed him to come in. To Zahara, Hassan was no more than a piece of furniture, and she thought as little of his intruding while she was in the midst of her toilet as another woman would have thought of the entrance of a maid.

"Two men," reported Hassan, "who won't go away until they see somebody."

"Whom do they want to see?" she inquired indifferently, adjusting the line of her eyebrow with an artistically pointed pencil.

"They say whoever belongs here."

Zahara invariably spoke either French or English to natives, and if Hassan had addressed her in Arabic she would not have replied, although she spoke that language better than she spoke any other.

"What are they like? Not--police?"

"Foreign," replied Hassan vaguely.

"English--American?"

"No, not American or English. Very black hair, dark skin."

Zahara, a student of men, became aware of a mild interest. These swarthy visitors should prove an agreeable antidote to the poisonous calm of Harry Grantham. She was trying with all the strength of her strange, stifled soul not to think of Grantham, and she was incapable of recognizing the fact that she could think of nothing else and had thought of little else for a long time past. Even now it was because of him that she determined to interview the foreign visitors. The mystery of her emotions puzzled her more than ever.

She descended to a small, barely furnished room on the ground floor, close beside the door opening upon the street. It was lighted by one hanging lamp. On the divan which constituted the principal item of furniture a small man, slenderly built, was sitting. He wore a broad-brimmed hat, so broad of brim that it threw the whole of the upper part of his face into shadow. It was impossible to see his eyes. Beside him rested a heavy walking-stick.

As Zahara entered, a wonderful, gaily colored figure, this man did not move in the slightest, but sat, chin on breast, his small, muscular, brown hands resting on his knees. His companion, however, a person of more massive build, elegantly dressed and handsome in a swarthy fashion, bowed gravely and removed his hat. Zahara liked his eyes, which were dark and very bold looking.

"M. Agapoulos is engaged," she said, speaking in French. "What is it you wish to know?"

The man regarded her fixedly, and:

"Senorita," he replied, "I will be frank with you."

Save for his use of the word "senorita" he also spoke in French. Zahara drew her robe more closely about her and adopted her most stately manner.

"My name," continued the other, "does not matter, but my business is to look into the affairs of other people, you understand?"

Zahara, who understood from this that the man was some kind of inquiry agent, opened her blue eyes very widely and at the same time shook her head.

"No," she protested; "what do you mean?"

"A certain gentleman came here a short time ago, came into this house and must be here now. Don't be afraid. He has done nothing very dreadful," he added reassuringly.

Zahara retreated a step, and a little wrinkle of disapproval appeared between her penciled brows. She no longer liked the man's eyes, she decided. They were deceitful eyes. His companion had taken up the heavy stick and was restlessly tapping the floor.

"There is no one here," said Zahara calmly, "except the people who live in the house."

"He is here, he is here," muttered the man seated on the divan.

The tapping of his stick had grown more rapid, but as he had spoken in Spanish, Zahara, who was ignorant of that language, had no idea what he had said.

"My friend," continued the Spaniard, bowing slightly in the direction of the slender man who so persistently kept his broad- brimmed hat on his head, "chanced to hear the voice of this gentleman as he spoke to your porter on entering the door. And although the door was closed too soon for us actually to see him, we are convinced that he is the person we seek."

"I think you are mistaken," said Zahara coolly. "But what do you want him for?"

As she uttered the words she realized that even the memory of Grantham was sufficient to cause her to betray herself. She had betrayed her interest to the man himself, and now she had betrayed it to this dark-faced stranger whose manner was so mysterious. The Spaniard recognized the fact, and, unlike Grantham, acted upon it promptly.

"He has taken away the wife of another, Senorita," he said simply, and watched her as he spoke the lie.

She listened in silence, wide-eyed. Her lower lip twitched, and she bit it fiercely.

"He went first to Port Said and then came to London with this woman," continued the Spaniard remorselessly. "We come from her husband to ask her to return. Yes, he will forgive her--or he offers her freedom."

Rapidly but comprehensively the speaker's bold glance traveled over Zahara, from her golden head to her tiny embroidered shoes.

"If you can help us in this matter it will be worth fifty English pounds to you," he concluded.

Zahara was breathing rapidly. The fatal hatred which she had sought to stifle gained a new vitality. Another woman--another woman actually here in London! So there was someone upon whom he did not look in that half-amused and half-compassionate manner. How she hated him! How she hated the woman to whom he had but a moment ago returned!

"Then he will marry this other one?" she said suddenly.

"Oh, no. Already he neglects her. We think she will go back."

Zahara experienced a swift change of sentiment. She seemed to be compounded of two separate persons, one of whom laughed cruelly at the folly of the other.

"What is the name of this man you think your friend has recognized?" she asked.

The big stick was rapping furiously during this colloquy.

"We are both sure, Senorita. His name is Major Spalding."

That Spalding and Grantham were neighboring towns in Lincolnshire Zahara did not know, but:

"No one of that name comes here," she replied.

"The one you heard and--who has gone--is not called by that name." She spoke with forced calm. It was Grantham they sought! "But what happens if I show you this one who is not called Spalding?"

"No matter! Point him out to me," answered the Spaniard eagerly --and his dark eyes seemed to be on fire--"point him out to me and fifty pounds of English money is yours!"

"Let me see."

He drew out a wallet and held up a number of notes.

"Fifty," he said, in a subdued voice, "when you point him out."

For a long moment Zahara hesitated, then:

"Sixty," she corrected him--"now! Then I will do it to-night--if you tell what happens."

Exhibiting a sort of eager impatience the man displayed a bunch of official-looking documents.

"I give him these," he explained, "and my work is done."

"H'm," said Zahara. "He must not know that it is I who have shown him to you. To-night he will be here at nine o'clock, and I shall dance. You understand?"

"Then," said the Spaniard eagerly, "this is what you will do."

And speaking close to her ear he rapidly outlined a plan; but presently she interrupted him.

"Pooh! It is Spanish, the rose. I dance the dances of Egypt."

"But to-night," he persisted, "it will not matter."

Awhile longer they talked, the rapping of the stick upon the tiled floor growing ever faster and faster. But finally:

"I will tell Hassan that you are to be admitted," said Zahara, and she held out her hand for the notes.

When, presently, the visitors departed, she learned that the smaller man was blind; for his companion led him out of the room and out of the house. She stood awhile listening to the tap, tap, tap of the heavy stick receding along the street. What she did not hear, and could not have understood had she heard, since it was uttered in Spanish, was the cry of exultant hatred which came from the lips of the taller man:

"At last, Miguel! at last! Though blind, you have found him! You have not failed. I shall not fail!"

*****

Zahara peeped through the carved screen at the assembled company. They were smoking and drinking and seemed to be in high good humor. Safiyeh had danced and they had applauded the performance, but had complained to M. Agapoulos that they had seen scores of such dances and dancers. Safiyeh, who had very little English, had not understood this, and because presently she was to play upon the a'ood while Zahara danced the Dance of the Veils, Zahara had avoided informing her of the verdict of the company.

Now as she peeped through the lattice in the screen she could see the Greek haggling with Grantham and a tall gray-haired man whom she supposed to be Sir Horace Tipton. They were debating the additional fees to be paid if Zahara, the Star of Egypt, was to present the secret and wonderful dance of which all men had heard but which only a true daughter of the ancient tribe of the Ghawazi could perform.

Sometimes Zahara was proud of her descent from a dancing-girl of Kenneh. This was always at night, when a sort of barbaric excitement possessed her which came from the blood of her mother. Then, a new light entered her eyes and they seemed to grow long and languid and dark, so that no one would have suspected that in daylight they were blue.

A wild pagan abandon claimed her, and she seemed to hear the wailing of reed instruments and the throb of the ancient drums which were played of old before the kings of Egypt. Safiyeh was not a true dancing girl, and because she knew none of those fine frenzies, she danced without inspiration, like a brown puppet moved by strings. But she could play upon an a'ood much better than Zahara, and therefore must not be upset until she had played for the Dance of the Veils.

Seeing that the bargain was all but concluded, Zahara stole back to her room. Her lightly clad body gleamed like that of some statue become animate.

Her cheeks flushed as she took up the veils, of which she alone knew the symbolic meaning; the white veil, the purple veil: each had its story to tell her; and the veil of burning scarlet. In a corner of the big room on a divan near the door she had seen the Spaniard, a handsome, swarthy figure in his well-fitting dress clothes, and now, opening a drawer, she glanced at the little pile of notes which represented her share of the bargain. There were fifty. She had told Agapoulos that a distinguished foreigner with an introduction from someone she knew had paid ten pounds to be present. And because she had given Agapoulos the ten pounds, Agapoulos had agreed to admit the visitor.

She could hear the Greek approaching now, but she was thinking of Grantham whom she had last seen in laughing conversation with the tall, gray-haired man. His laughter had appeared forced. Doubtless he grew weary of the woman he had brought to London.

"Dance to-night with all the devil that is in you, my beautiful," said Agapoulos, hurrying into the room.

Zahara turned aside, toying with the veils.

"They are rich, eh?" she said indifferently.

She was thinking of the fifty pounds which she had earned so easily; and after all (how strangely her mind wandered) perhaps he was really tired of the woman. The Spaniard had said so.

"Very rich," murmured Agapoulos complacently.

He brushed his moustache and rattled keys in his pocket. In his dress clothes he looked like the manager of a prosperous picture palace. "Safryeh!" he called.

When presently the music commenced, the players concealed behind the tall screen, an expectant hush fell upon the wine-flushed company. Hassan, who played the darabukkeh, could modulate its throbbing so wonderfully.

Zahara entered the room, enveloped from shoulders to ankles in a flame-colored cloak.

Between her lips she held a red rose.

"By God, what a beauty!" said a husky voice.

Zahara did not know which of the party had spoken, but she was conscious of the fact that by virtue of the strange witchcraft which became hers on such nights she held them all spell-bound. They were her slaves.

Slowly she walked across the apartment while the throbbing of the Arab drum grew softer and softer, producing a weird effect of space and distance. All eyes were fixed upon her, and meeting Grantham's gaze she saw at last the Light there which she knew. This sudden knowledge of triumph almost unnerved her, and the rose which she had taken from between her lips trembled in her white fingers. Two of the petals fell upon the carpet, which was cream-colored from the looms of Ispahan. Like blood spots the petals lay upon the cream surface.

Zahara swung sharply about. Agapoulos, seated alone in the chair over which he had draped the leopard skin, was busily brushing his moustache and glancing sideways toward the screen which concealed Safryeh. Zahara tilted her head on to her shoulder and cast a languorous glance into the shadows masking the watchful Spaniard.

She could see his eyes gleaming like those of a wild beast. An icy finger seemed to touch her heart. He had lied to her! She knew it, suddenly, intuitively. Well, she would see. She also had guile.

With a little scornful laugh Zahara tossed the rose on to the knees--of Agapoulos.

The sound of three revolver shots fired in quick succession rang out above the throbbing music. Agapoulos clutched at his shirt front with both hands, uttered a stifled scream and tried to stand up. He coughed, and glaring straight in front of him fell forward across a little coffee table laden with champagne bottles and glasses.

Coincident with the crash made by his falling body came the loud bang of a door. The Spaniard had gone.

"By God, sir! It's murder, it's murder!" cried the same husky voice which had commented upon the beauty of Zahara.

There was a mingling, purposeless movement. Someone ran to the door--to find that it was locked from the outside. Mr. Eddie, now recognizable by his accent, came toward the prone man, dazed, horrified, and grown very white. Zahara, a beautiful, tragic figure, in her flaming cloak, stood looking down at the dead man. Safiyeh was peeping round from behind the screen, her face a brown mask of terror. Hassan, holding his drum, appeared behind her, staring stupidly. To the smell of cigar smoke and perfume a new and acrid odor was added.

Vaguely the truth was stealing in upon the mind of the dancing-girl that she had been made party to a plot to murder Grantham. She had saved his life. He belonged to her now. She could hear him speaking, although for some reason she could not see him. A haze had come, blotting out everything but the still, ungainly figure which lay so near her upon the carpet, one clutching, fat hand, upon which a diamond glittered, outstretched so that it nearly touched her bare white feet.

148

"We must get out this way! The side door to the courtyard! None of us can afford to be mixed up in an affair of this sort."

There was more confused movement and a buzz of excited voices-- meaningless, chaotic. Zahara could feel the draught from the newly opened door. A thin stream of blood was stealing across the carpet. It had almost reached the fallen rose petals, which it strangely resembled in color under the light of the lanterns.

As though dispersed by the draught, the haze lifted, and Zahara saw Grantham standing by the open doorway through which he had ushered out the other visitors.

Wide-eyed and piteous she met his glance. She had seen that night the Look in his eyes. She had saved his life, and there was much, so much, that she wanted to tell him. A thousand yearnings, inexplicable, hitherto unknown, deep mysteries of her soul, looked out of those great eyes.

"Don't think," he said tensely, "that I was deceived. I saw the trick with the rose! You are as guilty as your villainous lover! Murderess!"

He went out and closed the door. The flame-colored cloak slowly slipped from Zahara's shoulders, and the veils, like falling petals, began to drop gently one by one upon the blood-stained carpet.

THE HAND OF THE MANDARIN QUONG

CHAPTER 1

THE SHADOW ON THE CURTAIN

"Singapore is by no means herself again," declared Jennings, looking about the lounge of the Hotel de l'Europe. "Don't you agree, Knox?"

Burton fixed his lazy stare upon the speaker.

"Don't blame poor old Singapore," he said. "There is no spot in this battered world that I have succeeded in discovering which is not changed for the worse."

Dr. Matheson flicked ash from his cigar and smiled in that peculiarly happy manner which characterizes a certain American type and which lent a boyish charm to his personality.

"You are a pair of pessimists," he pronounced. "For some reason best known to themselves Jennings and Knox have decided upon a Busman's Holiday. Very well. Why grumble?"

"You are quite right, Doctor," Jennings admitted. "When I was on service here in the Straits Settlements I declared heaven knows how often that the country would never see me again once I was demobbed. Yet here you see I am; Burton belongs here; but here's Knox, and we are all as fed up as we can be!"

"Yes," said Burton slowly. "I may be a bit tired of Singapore. It's a queer thing, though, that you fellows have drifted back here again. The call of the East is no fable. It's a call that one hears for ever."

The conversation drifted into another channel, and all sorts of topics were discussed, from racing to the latest feminine fashions, from ballroom dances to the merits and demerits of coalition government. Then suddenly:

"What became of Adderley?" asked Jennings.

There were several men in the party who had been cronies of ours during the time that we were stationed in Singapore, and at Jennings's words a sort of hush seemed to fall on those who had known Adderley. I cannot say if Jennings noticed this, but it was perfectly evident to me that Dr. Matheson had perceived it, for he glanced swiftly across in my direction in an oddly significant way.

"I don't know," replied Burton, who was an engineer. "He was rather an unsavory sort of character in some ways, but I heard that he came to a sticky end."

"What do you mean?" I asked with curiosity, for I myself had often wondered what had become of Adderley.

"Well, he was reported to his C. O., or something, wasn't he, just before the time for his

150

demobilization? I don't know the particulars; I thought perhaps you did, as he was in your regiment."

"I have heard nothing whatever about it," I replied.

"You mean Sidney Adderley, the man who was so indecently rich?" someone interjected. "Had a place at Katong, and was always talking about his father's millions?"

"That's the fellow."

"Yes," said Jennings, "there was some scandal, I know, but it was after my time here."

"Something about an old mandarin out Johore Bahru way, was it not?" asked Burton. "The last thing I heard about Adderley was that he had disappeared."

"Nobody would have cared much if he had," declared Jennings. "I know of several who would have been jolly glad. There was a lot of the brute about Adderley, apart from the fact that he had more money than was good for him. His culture was a veneer. It was his check-book that spoke all the time."

"Everybody would have forgiven Adderley his vulgarity," said Dr. Matheson, quietly, "if the man's heart had been in the right place."

"Surely an instance of trying to make a silk purse out of a sow's ear," someone murmured.

Burton gazed rather hard at the last speaker.

"So far as I am aware," he said, "the poor devil is dead, so go easy."

"Are you sure he is dead?" asked Dr. Matheson, glancing at Burton in that quizzical, amused way of his.

"No, I am not sure; I am merely speaking from hearsay. And now I come to think of it, the information was rather vague. But I gathered that he had vanished, at any rate, and remembering certain earlier episodes in his career, I was led to suppose that this vanishing meant..."

He shrugged his shoulders significantly.

"You mean the old mandarin?" suggested Dr. Matheson.

"Yes."

"Was there really anything in that story, or was it suggested by the unpleasant reputation of Adderley?" Jennings asked.

"I can settle any doubts upon that point," said I; whereupon I immediately became a focus of general attention.

"What! were you ever at that place of Adderley's at Katong?" asked Jennings with intense curiosity.

I nodded, lighting a fresh cigarette in a manner that may have been unduly leisurely.

"Did you see her?"

Again I nodded.

"Really!"

"I must have been peculiarly favored, but certainly I had that pleasure."

"You speak of seeing her" said one of the party, now entering the conversation for the first time. "To whom do you refer?"

"Well," replied Burton, "it's really a sort of fairy tale--unless Knox"--glacing across in my direction--"can confirm it. But there was a story current during the latter part of Adderley's stay in Singapore to the effect that he had made the acquaintance of the wife, or some member of the household, of an old gentleman out Johore Bahru way--sort of mandarin or big pot among the Chinks."

"It was rumored that he had bolted with her," added another speaker.

"I think it was more than a rumor."

"Why do you say so?"

"Well, representations were made to the authorities, I know for an absolute certainty, and I have an idea that Adderley was kicked out of the Service as a consequence of the scandal which resulted."

"How is it one never heard of this?"

"Money speaks, my dear fellow," cried Burton, "even when it is possessed by such a peculiar outsider as Adderley. The thing was hushed up. It was a very nasty business. But Knox was telling us that he had actually seen the lady. Please carry on, Knox, for I must admit that I am intensely curious."

"I can only say that I saw her on one occasion."

"With Adderley?"

"Undoubtedly."

"Where?"

"At his place at Katong."

"I even thought his place at that resort was something of a myth," declared Jennings. "He never asked me to go there, but, then, I took that as a compliment. Pardon the apparent innuendo, Knox," he added, laughing. "But you say you actually visited the establishment?"

"Yes," I replied slowly, "I met him here in this very hotel one evening in the winter of '15, after the natives' attempt to mutiny. He had been drinking rather heavily, a fact which he was quite unable to disguise. He was never by any means a real friend of mine; in fact, I doubt that he had a true friend in the world. Anyhow, I could see that he was lonely, and as I chanced to be at a loose end I accepted an invitation to go over to what he termed his 'little place at Katong.'

"His little place proved to be a veritable palace. The man privately, or rather, secretly, to be exact, kept up a sort of pagan state. He had any number of servants. Of course he became practically a millionaire after the death of his father, as you will remember; and given more congenial company, I must confess that I might have spent a most enjoyable evening there. "Adderley insisted upon priming me with champagne, and after a while I may as well admit that I lost something of my former reserve, and began in a fashion to feel that I was having a fairly good time. By the way, my host was not quite frankly drunk. He got into that objectionable and dangerous mood which some of you will recall, and I could see by the light in his eyes that there was mischief brewing, although at the time I did not know its nature.

"I should explain that we were amusing ourselves in a room which was nearly as large as the lounge of this hotel, and furnished in a somewhat similar manner. There were carved pillars and stained glass domes, a little fountain, and all those other peculiarities of an Eastern household.

"Presently, Adderley gave an order to one of his servants, and glanced at me with that sort of mocking, dare-devil look in his eyes which I loathed, which everybody loathed who ever met the man. Of course I had no idea what all this portended, but I was very shortly to learn.

"While he was still looking at me, but stealing side-glances at a doorway before which was draped a most wonderful curtain of a sort of flamingo color, this curtain was suddenly pulled aside, and a girl came in.

"Of course, you must remember that at the time of which I am speaking the scandal respecting the mandarin had not yet come to light. Consequently I had no idea who the girl could be. I saw she was a Eurasian. But of her striking beauty there could be no doubt whatever. She was dressed in magnificent robes, and she literally glittered with jewels. She even wore jewels upon the toes of her little bare feet. But the first thing that struck me at the moment of her appearance was that her presence there was contrary to her wishes and inclinations. I have never seen a similar expression in any woman's eyes. She looked at Adderley as though she would gladly have slain him!

"Seeing this look, his mocking smile in which there was something of triumph--of the joy of possession--turned to a scowl of positive brutality. He clenched his fists in a way that set me bristling. He advanced toward the girl--and although the width of the room divided them, she recoiled--and the significance of expression and gesture was unmistakable. Adderley paused.

"'So you have made up your mind to dance after all?' he shouted.

"The look in the girl's dark eyes was pitiful, and she turned to me with a glance of dumb entreaty.

"'No, no!' she cried. 'No, no! Why do you bring me here?'

"'Dance!' roared Adderley. 'Dance! That's what I want you to do.'

"Rebellion leapt again to the wonderful eyes, and she started back with a perfectly splendid gesture of defiance. At that my brutal and drunken host leapt in her direction. I was on my feet now, but before I could act the girl said a thing which checked him, sobered him, which pulled him up short, as though he had encountered a stone wall.

"'Ah, God!' she said. (She was speaking, of course, in her native tongue.) 'His hand! His hand! Look! His hand!'

"To me her words were meaningless, naturally, but following the direction of her positively agonized glance I saw that she was watching what seemed to me to be the shadow of someone moving behind the flame-like curtain which produced an effect not unlike that of a huge, outstretched hand, the fingers crooked, claw- fashion.

"'Knox, Knox!' whispered Adderley, grasping me by the shoulder.

"He pointed with a quivering finger toward this indistinct shadow upon the curtain, and:

"'Do you see it--do you see it?' he said huskily. 'It is his hand--it is his hand!'

"Of the pair, I think, the man was the more frightened. But the girl, uttering a frightful shriek, ran out of the room as though pursued by a demon. As she did so whoever had been moving behind the curtain evidently went away. The shadow disappeared, and Adderley, still staring as if hypnotized at the spot where it had been, continued to hold my shoulder as in a vise. Then, sinking down upon a heap of cushions beside me, he loudly and shakily ordered more champagne.

"Utterly mystified by the incident, I finally left him in a state of stupor, and returned to my quarters, wondering whether I had dreamed half of the episode or the whole of it, whether he did really possess that wonderful palace, or whether he had borrowed it to impress me."

I ceased speaking, and my story was received in absolute silence, until:

"And that is all you know?" said Burton.

"Absolutely all. I had to leave about that time, you remember, and afterward went to France."

"Yes, I remember. It was while you were away that the scandal arose respecting the mandarin. Extraordinary story, Knox. I should like to know what it all meant, and what the end of it was."

Dr. Matheson broke his long silence.

"Although I am afraid I cannot enlighten you respecting the end of the story," he said quietly, "perhaps I can carry it a step further."

"Really, Doctor? What do you know about the matter?"

"I accidentally became implicated as follows," replied the American: "I was, as you know, doing voluntary surgical work near Singapore at the time, and one evening, presumably about the same period of which Knox is speaking, I was returning from the hospital at Katong, at which I acted sometimes as anaesthetist, to my quarters in Singapore; just drifting along, leisurely by the edge of the gardens admiring the beauty of the mangroves and the deceitful peace of the Eastern night.

"The hour was fairly late and not a soul was about. Nothing disturbed the silence except those vague sibilant sounds which are so characteristic of the country. Presently, as I rambled on with my thoughts wandering back to the dim ages, I literally fell over a man who lay in the road.

"I was naturally startled, but I carried an electric pocket torch, and by its light I discovered that the person over whom I had fallen was a dignified-looking Chinaman, somewhat past middle age. His clothes, which were of good quality, were covered with dirt and blood, and he bore all the appearance of having recently been engaged in a very tough struggle. His face was notable only for its possession of an unusually long jet-black moustache. He had swooned from loss of blood."

"Why, was he wounded?" exclaimed Jennings.

"His hand had been nearly severed from his wrist!"

"Merciful heavens!"

"I realized the impossibility of carrying him so far as the hospital, and accordingly I extemporized a rough tourniquet and left him under a palm tree by the road until I obtained assistance. Later, at the hospital, following a consultation, we found it necessary to amputate."

"I should say he objected fiercely?"

"He was past objecting to anything, otherwise I have no doubt he would have objected furiously. The index finger of the injured hand had one of those preternaturally long nails, protected by an engraved golden case. However, at least I gave him a chance of life. He was under my care for some time, but I doubt if ever he was properly grateful. He had an iron constitution, though, and I finally allowed him to depart. One queer stipulation he had made--that the severed hand, with its golden nail-case, should be given to him when he left hospital. And this bargain I faithfully carried out."

"Most extraordinary," I said. "Did you ever learn the identity of the old gentleman?"

"He was very reticent, but I made a number of inquiries, and finally learned with absolute certainty, I think, that he was the Mandarin Quong Mi Su from Johore Bahru, a person of great repute among the Chinese there, and rather a big man in China. He was known locally as the Mandarin Quong."

"Did you learn anything respecting how he had come by his injury, Doctor?"

Matheson smiled in his quiet fashion, and selected a fresh cigar with great deliberation. Then:

"I suppose it is scarcely a case of betraying a professional secret," he said, "but during the time that my patient was recovering from the effects of the anaesthetic he unconsciously gave me several clues to the nature of the episode. Putting two and two together I gathered that someone, although the name of this person never once passed the lips of the mandarin, had abducted his favorite wife."

"Good heavens! truly amazing," I exclaimed.

"Is it not? How small a place the world is. My old mandarin had traced the abductor and presumably the girl to some house which I gathered to be in the neighborhood of Katong. In an attempt to force an entrance--doubtless with the amiable purpose of slaying them both--he had been detected by the prime object of his hatred. In hurriedly descending from a window he had been attacked by some weapon, possibly a sword, and had only made good his escape in the condition in which I found him. How far he had proceeded I cannot say, but I should imagine that the house to which he had been was no great

155

distance from the spot where I found him."

"Comment is really superfluous," remarked Burton. "He was looking for Adderley."

"I agree," said Jennings.

"And," I added, "it was evidently after this episode that I had the privilege of visiting that interesting establishment."

There was a short interval of silence; then:

"You probably retain no very clear impression of the shadow which you saw," said Dr. Matheson, with great deliberation. "At the time perhaps you had less occasion particularly to study it. But are you satisfied that it was really caused by someone moving behind the curtain?"

I considered his question for a few moments.

"I am not," I confessed. "Your story, Doctor, makes me wonder whether it may not have been due to something else."

"What else can it have been due to?" exclaimed Jennings contemptuously--"unless to the champagne?"

"I won't quote Shakespeare," said Dr. Matheson, smiling in his odd way. "The famous lines, though appropriate, are somewhat overworked. But I will quote Kipling: 'East is East, and West is West.'"

CHAPTER 2

THE LADY OF KATONG

Fully six months had elapsed, and on returning from Singapore I had forgotten all about Adderley and the unsavory stories connected with his reputation. Then, one evening as I was strolling aimlessly along St. James's Street, wondering how I was going to kill time--for almost everyone I knew was out of town, including Paul Harley, and London can be infinitely more lonely under such conditions than any desert--I saw a thick-set figure approaching along the other side of the street.

The swing of the shoulders, the aggressive turn of the head, were vaguely familiar, and while I was searching my memory and endeavoring to obtain a view of the man's face, he stared across in my direction.

It was Adderley.

He looked even more debauched than I remembered him, for whereas in Singapore he had had a tanned skin, now he looked unhealthily pallid and blotchy. He raised his hand, and:

"Knox!" he cried, and ran across to greet me.

His boisterous manner and a sort of coarse geniality which he possessed had made him popular with a certain set in former days, but I, who knew that this geniality was forced, and assumed to conceal a sort of appalling animalism, had never been deceived by it. Most people found Adderley out sooner or later, but I had detected the man's true nature from the very beginning. His eyes alone were danger signals for any amateur psychologist. However, I greeted him civilly enough:

"Bless my soul, you are looking as fit as a fiddle!" he cried. "Where have you been, and what have you been doing since I saw you last?"

"Nothing much," I replied, "beyond trying to settle down in a reformed world."

"Reformed world!" echoed Adderley. "More like a ruined world it has seemed to me."

He laughed loudly. That he had already explored several bottles was palpable.

We were silent for a while, mentally weighing one another up, as it were. Then:

"Are you living in town?" asked Adderley.

"I am staying at the Carlton at the moment," I replied. "My chambers are in the hands of the decorators. It's awkward. Interferes with my work."

"Work!" cried Adderley. "Work! It's a nasty word, Knox. Are you doing anything now?"

"Nothing, until eight o'clock, when I have an appointment."

"Come along to my place," he suggested, "and have a cup of tea, or a whisky and soda if you prefer it."

Probably I should have refused, but even as he spoke I was mentally translated to the lounge of the Hotel de l'Europe, and prompted by a very human curiosity I determined to accept his invitation. I wondered if Fate had thrown an opportunity in my way of learning the end of the peculiar story which had been related on that occasion.

I accompanied Adderley to his chambers, which were within a stone's throw of the spot where I had met him. That this gift for making himself unpopular with all and sundry, high and low, had not deserted him, was illustrated by the attitude of the liftman as we entered the hall of the chambers. He was barely civil to Adderley and even regarded myself with marked disfavor.

We were admitted by Adderley's man, whom I had not seen before, but who was some kind of foreigner, I think a Portuguese. It was characteristic of Adderley. No Englishman would ever serve him for long, and there had been more than one man in his old Company who had openly avowed his intention of dealing with Adderley on the first available occasion.

His chambers were ornately furnished; indeed, the room in which we sat more closely resembled a scene from an Oscar Asche production than a normal man's study. There was something unreal about it all. I have since thought that this unreality extended to the person of the man himself. Grossly material, he yet possessed an aura of mystery, mystery of an unsavory sort. There was something furtive, secretive, about Adderley's entire mode of life.

I had never felt at ease in his company, and now as I sat staring wonderingly at the strange

157

and costly ornaments with which the room was overladen I bethought me of the object of my visit. How I should have brought the conversation back to our Singapore days I know not, but a suitable opening was presently offered by Adderley himself.

"Do you ever see any of the old gang?" he inquired.

"I was in Singapore about six months ago," I replied, "and I met some of them again."

"What! Had they drifted back to the East after all?"

"Two or three of them were taking what Dr. Matheson described as a Busman's Holiday."

At mention of Dr. Matheson's name Adderley visibly started.

"So you know Matheson," he murmured. "I didn't know you had ever met him."

Plainly to hide his confusion he stood up, and crossing the room drew my attention to a rather fine silver bowl of early Persian ware. He was displaying its peculiar virtues and showing a certain acquaintance with his subject when he was interrupted. A door opened suddenly and a girl came in. Adderley put down the bowl and turned rapidly as I rose from my seat.

It was the lady of Katong!

I recognized her at once, although she wore a very up-to-date gown. While it did not suit her dark good looks so well as the native dress which she had worn at Singapore, yet it could not conceal the fact that in a barbaric way she was a very beautiful woman. On finding a visitor in the room she became covered with confusion.

"Oh," she said, speaking in Hindustani. "Why did you not tell me there was someone here?"

Adderley's reply was characteristically brutal.

"Get out," he said. "You fool."

I turned to go, for I was conscious of an intense desire to attack my host. But:

"Don't go, Knox, don't go!" he cried. "I am sorry, I am damned sorry, I..."

He paused, and looked at me in a queer sort of appealing way. The girl, her big eyes widely open, retreated again to the door, with curious lithe steps, characteristically Oriental. The door regained, she paused for a moment and extended one small hand in Adderley's direction.

"I hate you," she said slowly, "hate you! Hate you!"

She went out, quietly closing the door behind her. Adderley turned to me with an embarrassed laugh.

"I know you think I am a brute and an outsider," he said, "and perhaps I am. Everybody says I am, so I suppose there must be something in it. But if ever a man paid for his mistakes I have paid for mine, Knox. Good God, I haven't a friend in the world."

"You probably don't deserve one," I retorted.

"I know I don't, and that's the tragedy of it," he replied. "You may not believe it, Knox; I don't expect anybody to believe me; but for more than a year I have been walking on the edge of Hell. Do you know where I have been since I saw you last?"

I shook my head in answer.

"I have been half round the world, Knox, trying to find peace."

"You don't know where to look for it," I said.

"If only you knew," he whispered. "If only you knew," and sank down upon the settee, ruffling his hair with his hands and looking the picture of haggard misery. Seeing that I was still set upon departure:

"Hold on a bit, Knox," he implored. "Don't go yet. There is something I want to ask you, something very important."

He crossed to a sideboard and mixed himself a stiff whisky-and- soda. He asked me to join him, but I refused.

"Won't you sit down again?"

I shook my head.

"You came to my place at Katong once," he began abruptly. "I was damned drunk, I admit it. But something happened, do you remember?"

I nodded.

"This is what I want to ask you: Did you, or did you not, see that shadow?"

I stared him hard in the face.

"I remember the episode to which you refer," I replied. "I certainly saw a shadow."

"But what sort of shadow?"

"To me it seemed an indefinite, shapeless thing, as though caused by someone moving behind the curtain."

"It didn't look to you like--the shadow of a hand?"

"It might have been, but I could not be positive."

Adderley groaned.

"Knox," he said, "money is a curse. It has been a curse to me. If I have had my fun, God knows I have paid for it."

"Your idea of fun is probably a peculiar one," I said dryly.

Let me confess that I was only suffering the man's society because of an intense curiosity which now possessed me on learning that the lady of Katong was still in Adderley's company.

Whether my repugnance for his society would have enabled me to remain any longer I cannot say. But as if Fate had deliberately planned that I should become a witness of the concluding phases of this secret drama, we were now interrupted a second time, and again in a dramatic fashion.

Adderley's nondescript valet came in with letters and a rather large brown paper parcel sealed and fastened with great care.

As the man went out:

"Surely that is from Singapore," muttered Adderley, taking up the parcel.

He seemed to become temporarily oblivious of my presence, and his face grew even more haggard as he studied the writing upon the wrapper. With unsteady fingers he untied it, and I lingered, watching curiously. Presently out from the wrappings he took a very beautiful casket of ebony and ivory, cunningly carved and standing upon four claw-like ivory legs.

"What the devil's this?" he muttered.

He opened the box, which was lined with sandal-wood, and thereupon started back with a great cry, recoiling from the casket as though it had contained an adder. My former sentiments forgotten, I stepped forward and peered into the interior. Then I, in turn, recoiled.

In the box lay a shriveled yellow hand--with long tapering and well-manicured nails--neatly severed at the wrist!

The nail of the index finger was enclosed in a tiny, delicately fashioned case of gold, upon which were engraved a number of Chinese characters.

Adderley sank down again upon the settee.

"My God!" he whispered, "his hand! His hand! He has sent me his hand!"

He began laughing. Whereupon, since I could see that the man was practically hysterical because of his mysterious fears:

"Stop that," I said sharply. "Pull yourself together, Adderley. What the deuce is the matter with you?"

"Take it away!" he moaned, "take it away. Take the accursed thing away!"

"I admit it is an unpleasant gift to send to anybody," I said, "but probably you know more about it than I do."

"Take it away," he repeated. "Take it away, for God's sake, take it away, Knox!"

He was quite beyond reason, and therefore:

"Very well," I said, and wrapped the casket in the brown paper in which it had come. "What do you want me to do with it?"

"Throw it in the river," he answered. "Burn it. Do anything you like with it, but take it out of my sight!"

CHAPTER 3

THE GOLD-CASED NAIL

As I descended to the street the liftman regarded me in a curious and rather significant way. Finally, just as I was about to step out into the hall:

"Excuse me, sir," he said, having evidently decided that I was a fit person to converse with, "but are you a friend of Mr. Adderley's?"

"Why do you ask?"

"Well, sir, I hope you will excuse me, but at times I have thought the gentleman was just a little bit queer, like."

"You mean insane?" I asked sharply.

"Well, sir, I don't know, but he is always asking me if I can see shadows and things in the lift, and sometimes when he conies in late of a night he absolutely gives me the cold shivers, he does."

I lingered, the box under my arm, reluctant to obtain confidences from a servant, but at the same time keenly interested. Thus encouraged:

"Then there's that lady friend of his who is always coming here," the man continued. "She's haunted by shadows, too." He paused, watching me narrowly.

"There's nothing better in this world than a clean conscience, sir," he concluded.

*****

Having returned to my room at the hotel, I set down the mysterious parcel, surveying it with much disfavor. That it contained the hand of the Mandarin Quong I could not doubt, the hand which had been amputated by Dr. Matheson. Its appearance in that dramatic fashion confirmed Matheson's idea that the mandarin's injury had been received at the hands of Adderley. What did all this portend, unless that the Mandarin Quong was dead? And if he were dead why was Adderley more afraid of him dead than he had been of him living?

I thought of the haunting shadow, I thought of the night at Katong, and I thought of Dr. Matheson's words when he had told us of his discovery of the Chinaman lying in the road that night outside Singapore.

I felt strangely disinclined to touch the relic, and it was only after some moments' hesitation that I undid the wrappings and raised the lid of the casket. Dusk was very near and I had not yet lighted the lamps; therefore at first I doubted the evidence of my senses. But having lighted up and peered long and anxiously into the sandal-wood lining of the casket I could doubt no longer.

The casket was empty!

It was like a conjuring trick. That the hand had been in the box when I had taken it up from Adderley's table I could have sworn before any jury. When and by whom it had been removed was a puzzle beyond my powers of unraveling. I stepped toward the telephone-- and then remembered that Paul Harley was out of London. Vaguely wondering if Adderley had played me a particularly gruesome practical joke, I put the box on a sideboard and again contemplated the telephone doubtfully far a moment. It was in my mind to ring him up. Finally, taking all things into consideration, I determined that I would have nothing further to do with the man's unsavory and mysterious affairs.

It was in vain, however, that I endeavored to dismiss the matter from my mind; and throughout the evening, which I spent at a theatre with some American friends, I found myself constantly thinking of Adderley and the ivory casket, of the mandarin of Johore Bahru, and of the mystery of the shriveled yellow hand.

I had been back in my room about half an hour, I suppose, and it was long past midnight, when I was startled by a ringing of my telephone bell. I took up the receiver, and:

"Knox! Knox!" came a choking cry.

"Yes, who is speaking?"

"It is I, Adderley. For God's sake come round to my place at once!"

His words were scarcely intelligible. Undoubtedly he was in the grip of intense emotion.

"What do you mean? What is the matter?"

"It is here, Knox, it is here! It is knocking on the door! Knocking! Knocking!"

"You have been drinking," I said sternly. "Where is your man?"

"The cur has bolted. He bolted the moment he heard that damned knocking. I am all alone; I have no one else to appeal to." There came a choking sound, then: "My God, Knox, it is getting in! I can see. . . the shadow on the blind. . ."

Convinced that Adderley's secret fears had driven him mad, I nevertheless felt called upon to attend to his urgent call, and without a moment's delay I hurried around to St. James's Street. The liftman was not on duty, the lower hall was in darkness, but I raced up the stairs and found to my astonishment that Adderley's door was wide open.

"Adderley!" I cried. "Adderley!"

There was no reply, and without further ceremony I entered and searched the chambers. They were empty. Deeply mystified, I was about to go out again when there came a ring at the door-bell. I walked to the door and a policeman was standing upon the landing.

"Good evening, sir," he said, and then paused, staring at me curiously.

"Good evening, constable," I replied.

"You are not the gentleman who ran out awhile ago," he said, a note of suspicion coming into his voice.

I handed him my card and explained what had occurred, then:

"It must have been Mr. Adderley I saw," muttered the constable.

"You saw--when?"

"Just before you arrived, sir. He came racing out into St. James's Street and dashed off like a madman."

"In which direction was he going?"

"Toward Pall Mall."

*****

The neighborhood was practically deserted at that hour. But from the guard on duty before the palace we obtained our first evidence of Adderley's movements. He had raced by some five minutes before, frantically looking back over his shoulder and behaving like a man flying for his life. No one else had seen him. No one else ever did see him alive. At two o'clock there was no news, but I had informed Scotland Yard and official inquiries had been set afoot.

Nothing further came to light that night, but as all readers of the daily press will remember, Adderley's body was taken out of the pond in St. James's Park on the following day. Death was due to drowning, but his throat was greatly discolored as though it had been clutched in a fierce grip.

It was I who identified the body, and as many people will know, in spite of the closest inquiries, the mystery of Adderley's death has not been properly cleared up to this day. The identity of the lady who visited him at his chambers was never discovered. She completely disappeared.

The ebony and ivory casket lies on my table at this present moment, visible evidence of an invisible menace from which Adderley had fled around the world.

Doubtless the truth will never be known now. A significant discovery, however, was made some days after the recovery of Adderley's body.

From the bottom of the pond in St. James's Park a patient Scotland Yard official brought up the gold nail-case with its mysterious engravings--and it contained, torn at the root, the incredibly long finger-nail of the Mandarin Quong!

THE KEY OF THE TEMPLE OF HEAVEN

CHAPTER 1

THE KEEPER OF THE KEY

The note of a silver bell quivered musically through the scented air of the ante-room. Madame de Medici stirred slightly upon the divan with its many silken cushions, turning her head toward the closed door with the languorous, almost insolent, indifference which one perceives in the movements of a tigress. Below, in the lobby, where the pillars of Mokattam alabaster upheld the painted roof, the little yellow man from Pekin shivered slightly, although the air was warm for Limehouse, and always turned his mysterious eyes toward a corner of the great staircase which was visible from where he sat, coiled up, a lonely figure in the mushrabiyeh chair. Madame blew a wreath of smoke from her lips, and, through half-closed eyes, watched it ascend, unbroken, toward the canopy of cloth-of-gold which masked the ceiling. A Madonna by Leonardo da Vinci faced her across the apartment, the painted figure seeming to watch the living one upon the divan. Madame smiled into the eyes of the Madonna. Surely even the great Leonardo must have failed to reproduce that smile--the great Leonardo whose supreme art has captured the smile of Mona Lisa. Madame had the smile of Cleopatra, which, it is said, made Caesar mad, though in repose the beauty of Egypt's queen left him cold. A robe of Kashmiri silk, fine with a phantom fineness, draped her exquisite shape as the art of Cellini draped the classic figures which he wrought in gold and silver; it seemed incorporate with her beauty.

A second wreath of smoke curled upward to the canopy, and Madame watched this one also through the veil of her curved black lashes, as the Eastern woman watches the world through her veil. Those eyes were notable even in so lovely a setting, for they were of a hue rarely seen in human eyes, being like the eyes of a tigress; yet they could seem voluptuously soft, twin pools of liquid amber, in whose depths a man might lose his soul.

Again the silver bell sounded in the ante-room, and, below, the little yellow man shivered sympathetically. Again Madame stirred with that high disdain that so became her, who had the eyes of a tigress. Her carmine lips possessed the antique curve which we are told distinguished the lips of the Comtesse de Cagliostro; her cheeks had the freshness of flowers, and her hair the blackness of ebony, enhancing the miracle of her skin, which had the whiteness of ivory--not of African ivory, but of that fossil ivory which has lain for untold ages beneath the snows of Siberia.

She dropped the cigarette from her tapered fingers into a little silver bowl upon a table at her side, then lightly touched the bell which stood there also. Its soft note answered to the bell in the ante-room; a white-robed Chinese servant silently descended the great staircase, his soft red slippers sinking into the rich pile of the carpet; and the little yellow man from the great temple in Pekin followed him back up the stairway and was ushered into the presence of Madame de Medici.

164

The servant closed the door silently and the little yellow man, fixing his eyes upon the beautiful woman before him, fell upon his knees and bowed his forehead to the carpet.

Madame's lovely lips curved again in the disdainful smile, and she extended one bare ivory arm toward the visitor who knelt as a suppliant at her feet.

"Rise, my friend!" she said, in purest Chinese, which fell from her lips with the music of a crystal spring. "How may I serve you?"

The yellow man rose and advanced a step nearer to the divan, but the strange beauty of Madame had spoken straight to his Eastern heart, had, awakened his soul to a new life. His glance traveled over the vision before him, from the little Persian slipper that peeped below the drapery of Kashmir silk to the small classic head with its crown of ebon locks; yet he dared not meet the glance of the amber eyes.

"Sit here beside me," directed Madame, and she slightly changed her position with that languorous and lithe grace suggestive of a creature of the jungle.

Breathing rapidly betwixt the importance of his mission and a new, intoxicating emotion which had come upon him at the moment of entering the perfumed room, the yellow man obeyed, but always with glance averted from the taunting face of Madame. A golden incense-burner stood upon the floor, over between the high, draped windows, and a faint pencil from its dying fires stole greyly upward. Upon the scented smoke the Buddhist priest fixed his eyes, and began, with a rapidity that grew as he proceeded, to pour out his tale. Seated beside him, one round arm resting upon the cushions so as almost to touch him, Madame listened, watching the averted yellow face, and always smiling--smiling.

The tale was done at last; the incense-burner was cold, and breathlessly the Buddhist clutched his knees with lean, clawish fingers and swayed to and fro, striving to conquer the emotions that whirled and fought within him. Selecting another cigarette from the box beside her, and lighting it deliberately, Madame de Medici spoke.

"My friend of old," she said, and of the language of China she made strange music, "you come to me from your home in the secret city, because you know that I can serve you. It is enough."

She touched the bell upon the table, and the white-robed servant reentered, and, bowing low, held open the door. The little yellow man, first kneeling upon the carpet before the divan as before an altar, hurried from the apartment. As the door was reclosed, and Madame found herself alone again, she laughed lightly, as Calypso laughed when Ulysses' ship appeared off the shores of her isle.

God fashions few such women. It is well.

CHAPTER 2

THE TIGER LADY

"By heavens, Annesley!" whispered Rene Deacon, "what eyes that woman has!" His companion, following the direction of Deacon's glance, nodded rather grimly.

"The eyes of a Circe, or at times the eyes of a tigress."

"She is magnificent!" murmured Deacon rapturously. "I have never seen so beautiful a woman."

His glance followed the tall figure as it passed into a smaller salon on the left; nor was he alone in his regard. Fashionable society was well represented in the gallery--where a collection of pictures by a celebrated artist was being shown; and prior to the entrance of the lady in the strangely fashioned tiger-skin cloak, the somewhat extraordinary works of art had engaged the interest even of the most fickle, but, from the moment the tiger- lady made her appearance, even the most daring canvases were forgotten.

"She wears tiger-skin shoes!" whispered one.

"She is like a design for a poster!" laughed another.

"I have never seen anything so flashy in my life," was the acrid comment of a third.

"What a dazzlingly beautiful woman!" remarked another--this one a man. While:

"Who is she?" arose upon all sides.

Judging from the isolation of the barbaric figure, it would seem that society did not know the tiger-lady, but Deacon, seizing his companion by the arm and almost dragging him into the small salon which the lady had entered, turned in the doorway and looked into Annesley's eyes. Annesley palpably sought to evade the glance.

"You know everybody," whispered Deacon. "You must be acquainted with her."

A great number of people were now thronging into the room, not so much because of the pictures it contained, but rather out of curiosity respecting the beautiful unknown. Annesley tried to withdraw; his uneasiness grew momentarily greater.

"I scarcely know her well enough," he protested, "to present you. Moreover..."

"But she's smiling at you!" interrupted Deacon eagerly.

His handsome but rather weak face was flushed; he was, as an old clubman had recently said of him, "so very young." He lacked the restraint usual in cultured Englishmen, and had the frankly passionate manner which one associates with the South. His uncle, Colonel Deacon, a mordant wit, would say apologetically:

"Reggie" (Deacon's father) "married a Gascon woman. She was delightfully pretty. Poor Reggie!"

Certainly Rene was impetuous to an embarrassing degree, nor lightly to be thwarted. Boldly meeting the glance of the woman of the amber eyes, he pushed Annesley forward, not troubling to disguise his anxiety to be presented to the tiger-lady. She turned her head languidly, with that wild-animal grace of hers, and unsmiling now, regarded Annesley.

"So you forget me so soon, Mr. Annesley," she murmured, "or is it that you play the good shepherd?"

"My dear Madame," said Annesley, recovering with an effort his wonted sang-froid, "I was merely endeavoring to calm the rhapsodies of my friend, who seemed disposed to throw himself at your feet in knight-errant fashion."

"He is a very handsome boy," murmured Madame; and as the great eyes were turned upon Deacon the carmine lips curved again in the Cleopatrian smile.

She was indeed wonderful, for while she spoke as the woman of the world to the boy, there was nothing maternal in her patronage, and her eyes were twin flambeaux, luring-- luring, and her sweet voice was a siren's song.

"May I beg leave to present my friend, Mr. Rene Deacon, Madame de Medici?" said Annesley; and as the two exchanged glances--the boy's a glance of undisguised passionate admiration, the woman's a glance unfathomable--he slightly shrugged his shoulders and stood aside.

There were others in the salon, who, perceiving that the unknown beauty was acquainted with Annesley, began to move from canvas to canvas toward that end of the room where the trio stood. But Madame did not appear anxious to make new acquaintances.

"I have seen quite enough of this very entertaining exhibition," she said languidly, toying with a great unset emerald which swung by a thin gold chain about her neck. "Might I entreat you to take pity upon a very lonely woman and return with me to tea?"

Annesley seemed on the point of refusing, when:

"I have acquired a reputed Leonardo," continued Madame, "and I wish you to see it."

There was something so like a command in the words that Deacon stared at his companion in frank surprise. The latter avoided his glance, and:

"Come!" said Madame de Medici.

As of old the great Catherine of her name might have withdrawn with her suite, so now the lady of the tiger skins withdrew from the gallery, the two men following obediently, and one of them at least a happy courtier.

CHAPTER 3

TWIN POOLS OF AMBER

THE white-robed Chinese servant entered and placed fresh perfume upon the burning charcoal of the silver incense-burner. As the scented smoke began to rise he withdrew, and a second servant entered, who facially, in dress, in figure and bearing, was a duplicate of the first. This one carried a large tray upon which was set an exquisite porcelain tea- service. He placed the tray upon a low table beside the divan, and in turn withdrew.

Deacon, seated in a great ebony chair, smoked rapidly and nervously--looking about the strangely appointed room with its huge picture of the Madonna, its jade Buddha surmounting a gilded Burmese cabinet, its Persian canopy and Egyptian divan, at the thousand and one costly curiosities which it displayed, at this mingling of East and West, of

167

Christianity and paganism, with a growing wonder.

To one of his blood there was delight, intoxication, in that room; but something of apprehension, too, now grew up within him.

Madame de Medici entered. The garish motor-coat was discarded now, and her supple figure was seen to best advantage in one of those dark silken gowns which she affected, and which had a seeming of the ultra-fashionable because they defied fashion. She held in her hand an orchid, its structure that of an odontoglossum, but of a delicate green color heavily splashed with scarlet--a weird and unnatural-looking bloom.

Just within the doorway she paused, as Deacon leaped up, and looked at him through the veil of the curved lashes.

"For you," she said, twirling the blossom between her fingers and gliding toward him with her tigerish step.

He spoke no word, but, face flushed, sought to look into her eyes as she pinned the orchid in the button-hole of his coat. Her hands were flawless in shape and coloring, being beautiful as the sculptured hands preserved in the works of Phidias.

The slight draught occasioned by the opening of the door caused the smoke from the incense-burner to be wafted toward the centre of the room. Like a blue-gray phantom it coiled about the two standing there upon a red and gold Bedouin rug, and the heavy perfume, or the close proximity of this singularly lovely woman, wrought upon the high-strung sensibilities of Deacon to such an extent that he was conscious of a growing faintness.

"Ah! You are not well!" exclaimed Madame with deep concern. "It is the perfume which that foolish Ah Li has lighted. He forgets that we are in England."

"Not at all," protested Deacon faintly, and conscious that he was making a fool of himself. "I think I have perhaps been overdoing it rather of late. Forgive me if I sit down."

He sank on the cushioned divan, his heart beating furiously, while Madame touched the little bell, whereupon one of the servants entered.

She spoke in Chinese, pointing to the incense-burner.

Ah Li bowed and removed the censer. As the door softly reclosed:

"You are better?" she whispered, sweetly solicitous, and, seating herself beside Deacon, she laid her hand lightly upon his arm.

"Quite," he replied hoarsely; "please do not worry about me. I am wondering what has become of Annesley."

"Ah, the poor man!" exclaimed Madame, with a silver laugh, and began to busy herself with the teacups. "He remembered, as he was looking at my new Leonardo, an appointment which he had quite forgotten."

"I can understand his forgetting anything under the circumstances."

Madame de Medici raised a tiny cup and bent slightly toward him. He felt that he was losing control of himself, and, averting his eyes, he stooped and smelled the orchid in his buttonhole. Then, accepting the cup, he was about to utter some light commonplace when the faintness returned overwhelmingly, and, hurriedly replacing the cup upon the tray, he fell back among the cushions. The stifling perfume of the place seemed to be choking him.

"Ah, poor boy! You are really not at all well. How sorry I am!"

The sweet tones reached him as from a great distance; but as one dying in the desert turns his face toward the distant oasis, Deacon turned weakly to the speaker. She placed one fair arm behind his head, pillowing him, and with a peacock fan which had lain amid the cushions fanned his face. The strange scene became wholly unreal to him; he thought himself some dying barbaric chief.

"Rest there," murmured the sweet voice.

The great eyes, unveiled now by the black lashes, were two twin lakes of fairest amber. They seemed to merge together, so that he stood upon the brink of an unfathomable amber pool--which swallowed him up--which swallowed him up.

He awoke to an instantaneous consciousness of the fact that he had been guilty of inexcusably bad form. He could not account for his faintness, and reclining there amid the silken cushions, with Madame de Medici watching him anxiously, he felt a hot flush stealing over his face.

"What is the matter with me!" he exclaimed, and sprang to his feet. "I feel quite well now."

She watched him, smiling, but did not speak. He was a "very young man" again, and badly embarrassed. He glanced at his wrist-watch.

"Gracious heavens!" he cried, and noted that the tea-tray had been removed, "there must be something radically wrong with my health. It is nearly seven o'clock!"

The note of the silver bell sounded in the ante-room.

"Can you forgive me?" he said.

But Madame, rising to her feet, leaned lightly upon his shoulder, toying with the petals of the orchid in his buttonhole.

"I think it was the perfume which that foolish Ah Li lighted," she whispered, looking intently into his eyes, "and it is you who have to forgive me. But you will, I know!" The silver bell rang again. "When you have come to see me again--many, many times, you will grow to love it--because I love it."

She touched the bell upon the table, and Ah Li entered silently. When Madame de Medici held out her hand to him Deacon raised the white fingers to his lips and kissed them rapturously; then he turned, the Gascon within him uppermost again, and ran from the room.

A purple curtain was drawn across the lobby, screening the caller newly arrived from the one so hurriedly departing.

CHAPTER 4

THE LIVING BUDDHA

It was past midnight when Colonel Deacon returned to the house. Rene was waiting for him, pacing up and down the big library. Their relationship was curious, as subsisting between ward and guardian, for these two, despite the disparity of their ages, had few secrets from one another. Rene burned to pour out his story of the wonderful Madame de Medici, of the secret house in Chinatown with its deceptively mean exterior and its gorgeous interior, to the shrewd and worldly elder man. That was his way. But Fate had an oddly bitter moment in store for him.

"Hallo, boy!" cried the Colonel, looking into the library; "glad you're home. I might not see you in the morning, and I want to tell you about--er--a lady who will be coming here in the afternoon."

The words died upon Rene's lips unspoken, and he stared blankly at the Colonel.

"I thought I knew all there was to know about pictures, antiques, and all that sort of lumber," continued Colonel Deacon in his rapid and off-hand manner. "Thought there weren't many men in London could teach me anything; certainly never suspected a woman could. But I've met one, boy! Gad! What a splendid creature! You know there isn't much in the world I haven't seen--north, south, east and west. I know all the advertised beauties of Europe and Asia--stage, opera, and ballet, and all the rest of them. But this one--Gad!"

He dropped into an arm-chair, clapping both his hands upon his knees. Rene stood at the farther end of the library, in the shadow, watching him.

"She's coming here to-morrow, boy--coming here. Gad! you dog! You'll fall in love with her the moment you see her--sure to, sure to! I did, and I'm three times your age!"

"Who is this lady, sir?" asked Rene, very quietly.

"God knows, boy! Everybody's mad to meet her, but nobody knows who she is. But wait till you see her. Lady Dascot seems to be acquainted with her, but you will see when they come to-morrow-- see for yourself. Gad, boy! . . . what did you say?"

"I did not speak."

"Thought you did. Have a whisky-and-soda ?"

"No, thank you, sir--good night."

"Good night, boy!" cried the Colonel. "Good night. Don't forget to be in to-morrow afternoon or you'll miss meeting the loveliest woman in London, and the most brilliant."

"What is her name?"

"Eh? She calls herself Madame de Medici. She's a mystery, but what a splendid creature!"

Rene Deacon walked slowly upstairs, entered his bedroom, and for fully an hour sat in the darkness, thinking--thinking.

"Am I going mad?" he murmured. "Or is this witch driving all London mad?"

He strove to recover something of the glamour which had mastered him when in the presence of Madame de Medici, but failed. Yet he knew that, once near her again, it would all return. His reflections were bitter, and when at last wearily he undressed and went to bed it was to toss restlessly far into the small hours ere sleep came to soothe his troubled mind.

But his sleep was disturbed: a series of dreadfully realistic dreams danced through his brain. First he seemed to be standing upon a high mountain peak with eternal snows stretched all about him. He looked down, past the snow line, past the fir woods, into the depths of a lovely lake, far down in the valley below. It was a lake of liquid amber, and as he looked it seemed to become two lakes, and they were like two great eyes looking up at him and summoning him to leap. He thought that he leaped, a prodigious leap, far out into space; then fell--fell--fell. When he splashed into the amber deeps they became churned up in a milky foam, and this closed about him with a strangle grip. But it was no longer foam, but the clinging arms of Madame de Medici! . . .

Then he stood upon a fragile bridge of bamboo spanning a raging torrent. Right and left of the torrent below were jungles in which moved tigerish shapes. Upon the farther side of the bridge Madame de Medici, clad in a single garment of flame-colored silk, beckoned to him. He sought to cross the bridge, but it collapsed, and he fell near the edge of the torrent. Below were the raging waters, and ever nearing him the tigerish shapes, which now Madame was calling to as to a pack of hounds. They were about to devour him, when...

He was crouching upon a ledge, high above a street which seemed to be vaguely familiar. He could not see very well, because of a silk mask tied upon his face, and the eyeholes of which were badly cut. From the ledge he stepped to another, perilously. He gained it, and crouching there, where there was scarce foothold for a cat, he managed fully to raise a window which already was raised some six inches. Then softly and silently--for he was bare-footed--he entered the room.

Someone slept in a bed facing the window by which he had entered, and upon a table at the side of the sleeper lay a purse, a bunch of keys, an electric torch, and a Service revolver. Gliding to the table Rene took the keys and the electric torch, unlocked the door of the room, and crept down a thickly carpeted stair to a room below. The door of this also he opened with one of the keys in the bunch, and by the light of the torch found his way through a quantity of antique furniture and piled up curiosities to a safe set in the farther wall.

He seemed, in his dream, to be familiar with the lock combination, and, selecting the correct key from the bunch, he soon had the safe open. The shelves within were laden principally with antique jewellery, statuettes, medals, scarabs; and a number of little leather-covered boxes were there also. One of these he abstracted, relocked the safe, and stepped out of the room, locking the door behind him. Up the stairs he mounted to the bedroom wherein he had left the sleeper. Having entered, he locked the door from within, placed the keys and the torch upon the table, and crept out again upon the dizzy ledge.

Poised there, high above the thoroughfare below, a great nausea attacked him. Glancing to the right, in the direction of the window through which he had come, he perceived

171

Madame de Medici leaning out and beckoning to him. Her arm gleamed whitely in the faint light. A new courage came to him. He succeeded, crouched there upon the narrow ledge, in relowering the window, and leaving it in the state in which he had found it, he stood up and essayed that sickly stride to the adjoining ledge. He accomplished it, knelt, and crept back into the room from which he had started. . . .

The head of an ivory image of Buddha loomed up out of the utter darkness, growing and growing until it seemed like a great mountain. He could not believe that there was so much ivory in the world, and he felt it with his fingers, wonderingly. As he did so it began to shrink, and shrink, and shrink, and shrink, until it was no larger that a seated human figure. Then beneath his trembling hands it became animate; it moved, extended ivory arms, and wrapped them about his neck. Its lips became carmine-- perfumed; they bent to him. . . and he was looking into the bewitching face of Madame de Medici!

He awoke, gasping for air and bathed in cold perspiration. The dawn was just breaking over London and stealing greyly from object to object in his bedroom.

CHAPTER 5

THE IVORY GOD

The great car, with its fittings of gold and ivory, drew up at the door of Colonel Deacon's house. The interior was ablaze with tiger lilies, and out from their midst stepped the fairest of them all--Madame de Medici, and swept queenly up the steps upon the arm of the cavalierly soldier.

All connoisseurs esteemed it a privilege to view the Deacon collection, and this afternoon there was a goodly gathering. Chairs and little white tables were dotted about the lawn in shady spots, and the majority of the company were already assembled; but when, in a wonderful golden robe, Madame de Medici glided across the lawn, the babel ceased abruptly as if by magic. She pulled off one glove and began twirling a great emerald between her slim fingers. It was suspended from a thin gold chain. Presently, descrying Annesley seated at a table with Lady Dascot, she raised the jewel languidly and peered through it at the two.

"Why!" exclaimed Rene Deacon, who stood close beside her, "that was a trick of Nero's!"

Madame laughed musically.

"One might take a worse model," she said softly; "at least he enjoyed life."

Colonel Deacon, who listened to her every word as to the utterance of a Cumaean oracle, laughed with extraordinary approbation.

There was scarce a woman present who regarded Madame with a friendly eye, nor a man who did not aspire to become her devoted slave. She brought an atmosphere of unreality with her, dominating old and young alike by virtue of her splendid pagan beauty. The lawn, with its very modern appointments, became as some garden of the Golden House, a pleasure ground of an emperor.

But later, when the company entered the house, and Colonel Deacon sought to monopolize the society of Madame, an unhealthy spirit of jealousy arose between Rene and his guardian. It was strange, grotesque, horrible almost. Annesley watched from afar, and there was something very like anger in his glance.

"And this," said the Colonel presently, taking up an exquisitely carved ivory Buddha, "has a strange history. In some way a legend has grown up around it--it is of very great age--to the effect that it must always cause its owner to lose his most cherished possession."

"I wonder," said the silvern voice, "that you, who possess so many beautiful things, should consent to have so ill-omened a curiosity in your house."

"I do not fear the evil charm of this little ivory image," said Colonel Deacon, "although its history goes far to bear out the truth of the legend. Its last possessor lost his most cherished possession a month after the Buddha came into his hands. He fell down his own stairs-- and lost his life!"

Madame de Medici languidly surveyed the figure through the upraised emerald.

"Really!" she murmured. "And the one from whom he procured it?"

"A Hindu usurer of Simla," replied the Colonel. "His daughter stole it from her father together with many other things, and took them to her lover, with whom she fled!"

Madame de Medici seemed to be slightly interested.

"I should love to possess so weird a thing," she said softly.

"It is yours!" exclaimed the Colonel, and placed it in her hands.

"Oh, but really," she protested.

"But really I insist--in order that you may not forget your first visit to my house!"

She shrugged her shoulders.

"How very kind you are, Colonel Deacon," she said, "to a rival collector!"

"Now that the menace is removed," said Colonel Deacon with labored humor, "I will show you my most treasured possession."

"So! I am greatly interested."

"Not even this rascal Rene," said the Colonel, stopping before a safe set in the wall, "has seen what I am about to show you!"

Rene started slightly and watched with intense interest the unlocking of the safe.

"If I am not superstitious about the ivory Buddha," continued the Colonel, "I must plead guilty in the case of the Key of the Temple of Heaven!"

"The Key of the Temple of Heaven!" murmured a lady standing immediately behind Madame de Medici. "And what is the Key of the Temple of Heaven?"

The Colonel, having unlocked the safe, straightened himself, and while everyone was waiting to see what he had to show, began to speak again pompously:

173

"The Temple of Heaven stands in the outer or Chinese City of Pekin, and is fabulously wealthy. No European, I can swear, had ever entered its secret chambers until last year. One of its most famous treasures was this Key. It was used only to open the special entrance reserved for the Emperor when he came to worship after his succession to the throne--that was, of course, before China became a Republic. The Key is studded almost all over with precious stones. Last year a certain naval man--I'll not mention his name-- discovered the secret of its hiding-place. How he came by that knowledge does not matter at present. One very dark night he crept up to the temple. He found the Keeper of the Key- - a Buddhist priest--to be sleeping, and he succeeded, therefore, in gaining access and becoming possessed of the Key."

A chorus of excited exclamations greeted this dramatic point of the story.

"The object of this outrage," continued the Colonel, "for an outrage I cannot deny it to have been, was not a romantic one. The poor chap wanted money, and he thought he could sell the Key to one of the native jewelers. But he was mistaken. He got back safely, and secretly offered it in various directions. No one would touch the thing; moreover, although of great value, the stones were very far from flawless, and not really worth the risks which he had run to secure them. Don't misunderstand me; the Key would fetch a big sum, but not a fortune."

"Yes?" said Madame de Medici, smiling, for the Colonel paused.

"He packed it up and addressed it to me, together with a letter. The price that he asked was quite a moderate one, and when the Key arrived in England I dispatched a check immediately. It never reached him."

"Why?" cried many whom this strange story had profoundly interested.

"He was found dead at the back of the native cantonments, with a knife in his heart!"

"Oh!" exclaimed Lady Dascot. "How positively ghastly! I don't think I want to see the dreadful thing!"

"Really!" murmured Madame de Medici, turning languidly to the speaker. "I do."

The Colonel stooped and reached into the safe. Then he began to take out object after object, box after box. Finally, he straightened himself again, and all saw that his face was oddly blanched.

"It's gone!" he whispered hoarsely. "The Key of the Temple of Heaven has been stolen!"

CHAPTER 6

MADAME SMILES

Rene entered his bedroom, locked the door, and seated himself on the bed; then he lowered his head into his hands and clutched at his hair distractedly. Since, on his uncle's own showing, no one knew that the Key of the Temple of Heaven had been in the safe, since,

174

excepting himself (Rene) and the Colonel, no one else knew the lock combination, how the Key had been stolen was a mystery which defied conjecture. No one but the Colonel had approached within several yards of the safe at the time it was opened; so that clearly the theft had been committed prior to that time.

Now Rene sought to recall the details of a strange dream which he had dreamed immediately before awakening on the previous night; but he sought in vain. His memory could supply only blurred images. There had been a safe in his dream, and he--was it he or another?--had unlocked it. Also there had been an enormous ivory Buddha. . . . Yet, stay! it had not been enormous; it had been. . .

He groaned at his own impotency to recall the circumstances of that mysterious, perhaps prophetic dream; then in despair he gave it up, and stooping to a little secretaire, unlocked it with the idea of sending a note round to Annesley's chambers. As he did so he uttered a loud cry.

Lying in one of the pigeon-holes was a long piece of black silk, apparently torn from the lining of an opera hat. In it two holes were cut as if it were intended to be used as a mask. Beside it lay a little leather-covered box. He snatched it out and opened it. It was empty!

"Am I going mad?" he groaned. "Or..."

"You are wanted on the 'phone, sir."

It was the butler who had interrupted him. Rene descended to the telephone, dazedly, but, recognizing the voice of Annesley, roused himself.

"I'm leaving town to-night, Deacon," said Annesley, "for--well, many reasons. But before I go I must give you a warning, though I rely on you never to mention my name in the matter. Avoid the woman who calls herself Madame de Medici; she'll break you. She's an adventuress, and has a dangerous acquaintance with Eastern cults, and. . . I can't explain properly. . . ."

"Annesley! the Key!"

"It's the theft of the Key that has prompted me to speak, Deacon. Madame has some sort of power--hypnotic power. She employed it on me once, to my cost! Paul Harley, of Chancery Lane, can tell you more about her. The house she's living in temporarily used to belong to a notorious Eurasian, Zani Chada. To make a clean breast of it I daren't thwart her openly; but I felt it up to me to tell you that she possesses the secret of post-hypnotic suggestion. I may be wrong, but I think you stole that Key!"

"I!"

"She hypnotized you at some time, and, by means of this uncanny power of hers, ordered you to steal the Key of the Temple of Heaven in such and such a fashion at a certain hour in the night. . ."

"I had a strange seizure while I was at her house. . . ."

"Exactly! During that time you were receiving your hypnotic orders. You would remember nothing of them until the time to execute them--which would probably be during sleep. In a state of artificial somnambulism, and under the direction of Madame's will, you became a burglar!"

As Madame de Medici's car drove off from the house of Colonel Deacon, and Madame seated herself in the cushioned corner, up from amid the furs upon the floor, where, dog-like, he had lain concealed, rose the little yellow man from the Temple of Heaven. He extended eager hands toward her, kneeling there, and spoke:

"Quick! quick!" he breathed. "You have it? The Key of the Temple."

Madame held in her hand an ivory Buddha. Inverting it she unscrewed the pedestal, and out from the hollow inside the image dropped a gleaming Key.

"Ah!" breathed the yellow man, and would have clutched it; but Madame disdainfully raised her right hand which held the treasure, and with her left hand thrust down the clutching yellow fingers.

She dropped the Key between her white skin and the bodice of her gown, tossing the ivory figure contemptuously amid the fur.

"Ah!" repeated the yellow man in a different tone, and his eyes gleamed with the flame of fanaticism. He slowly uprose, a sinister figure, and with distended fingers prepared to seize Madame by the throat. His eyes were bloodshot, his nostrils were dilated, and his teeth were exposed like the fangs of a wolf.

But she pulled off her glove and stretched out her bare white hand to him as a queen to a subject; she raised the long curved lashes, and the great amber eyes looked into the angry bloodshot eyes.

The little yellow man began to breathe more and more rapidly; soon he was panting like one in a fight to the death who is all but conquered. At last he dropped on his knees amid the fur. . . and the curling lashes were lowered again over the blazing amber eyes that had conquered.

Madame de Medici lowered her beautiful white hand, and the little yellow man seized it in both his own and showered rapturous kisses upon it.

Madame smiled slightly.

"Poor little yellow man!" she murmured in sibilant Chinese, "you shall never return to the Temple of Heaven!"

www.ingramcontent.com/pod-product-compliance
Lightning Source LLC
Chambersburg PA
CBHW051411200326
41520CB00023B/7193